St. An   n

f

Lough Gill, Sligo.
Telephone (071) 43580 or 42785

D1320802

MICRO-ORGANISMS

576

This book is due for return on or before the last date shown below.

		0 2 DEC 1997
		07 NOV 2001
		2 NOV 2001
		8 OCT 2001
		06 JAN 2005
		07 NOV 2005

Don Gresswell Ltd., London, N.21   Cat. No. 1208

DG 02242/71

MICRO-ORGANISMS

by

J. I. WILLIAMS, B.Sc., Ph.D., Dip. Ed., M.I. Biol.
Senior Lecturer in Biology, People's College of
Further Education, Nottingham

and

M. SHAW, M.Sc., M.I. Biol.

Senior Lecturer in Biology, Stockport College of
Technology

Series Editor
M. K. Sands
Lecturer in Education, University of Nottingham

First published in Great Britain 1976 by Mills & Boon Limited,
17–19 Foley Street, London W1A 1DR.

© J. I. Williams and M. Shaw 1976

Reprinted 1978

ISBN 0 263 06104 3

Printed in Great Britain by
Thomson Litho Limited, East Kilbride, Scotland

CONTENTS

ACKNOWLEDGEMENTS

We gratefully acknowledge the assistance of:

Our editor, Margaret K. Sands, of the University of Nottingham, for her invaluable advice and considerable patience during the preparation of the text.

Professor G. J. F. Pugh, University of Aston, for reading the manuscript.

The Journal of Bacteriology, and Dr. Edgar Ribi for providing Plate 2.1.

Dr. J. Freer, University of Glasgow, for Plates 2.2, 2.3, 2.5, 2.6, 5.1 (1 and 2) and 5.2.

Heinemann Educational Books and Mr. B. Bracegirdle for Plate 4.1.

Dr. G. F. Leedale, University of Leeds, for Plate 4.2.

Dr. R. K. Gibson, University of Manchester, for Plates 6.2, 6.3 and 6.4.

The Forestry Commission for Plate 6.8.

Mr. Peter Fry for Plate 8.1.

Dr. R. Coutts, University of Birmingham, for Plates 5.1, (3) and 8.5.

Imperial Chemical Industries Limited for Figs. 8.8 and 8.9.

Cambridge University Press, *Transactions of the British Mycological Society* and Mr. A. J. P. Trinci for permission to use a graph as a basis for Figure 7.3.

Robinson's Brewery, Stockport, for Plate 8.2.

Glaxo Laboratories Ltd., for Plate 8.3.

The Joint Matriculation Board for giving permission to use a number of JMB and Nuffield 'A' level examination questions.

Any errors in the text are entirely the responsibility of the authors.

PREFACE

The content and scope of biological courses are forever being revised and there is now a trend towards the emphasis of practical applications of biological knowledge. Micro-organisms exert profound effects on a wide range of life processes so they are extremely important to man. Microbiological activities such as fermentation by yeasts were used by man for thousands of years but during the past century there has been a quickening of interest in micro-organisms. This is in no small way due to the sound foundations of microbial knowledge established by Pasteur and Koch in the latter part of the nineteenth century. Such has been the advance in our understanding of these organisms that vast industries are now centred around their activities.

This book was written to provide a text for the optional topic on MICRO-ORGANISMS in the new Joint Matriculation Board G.C.E. 'A' level syllabus in Biology. We also hope that it will find a place in many schools and colleges where other biological courses are followed. There is much fundamental matter on microbes in the earlier pages, whilst some of the practical exercises can be employed to investigate basic biological phenomena. Omission of the algae is regretted but we felt that we should concentrate, in a book of this size, on the organisms traditionally accepted as microbes.

Although we could have recommended other more sophisticated techniques in the practical section we have selected methods which require simple equipment available in most schools and colleges. A section on safety is included which should be read before beginning any practical work.

1 WHAT ARE MICRO-ORGANISMS?

Micro-organisms are generally thought of as those organisms whose small size makes it impossible for them to be seen clearly without the aid of a microscope. They include protozoa which are unmistakably animals, algae which are clearly plants, and other organisms such as fungi and bacteria which are not obviously plants or animals. Some fungi are even macroscopic.

As long ago as 1866, the German biologist Ernst Haeckel proposed a new kingdom, the *Protista,* to contain the large number of micro-organisms which were observed for the first time as methods of investigation were improved. His ideas were generally rejected by his colleagues who regarded these small forms as either simple plants or simple animals. The best we can say is that each of the main protistan life forms,

(*a*) viruses	(*b*) blue–green algae	(*c*) bacteria
(*d*) fungi	(*e*) algae	(*f*) protozoa

contain organisms able to carry out living processes common to microbe and man alike. An important step in studying the *Protista* was the realization that, of these so-called simple forms, some appeared more simple than others. In 1962 Stanier and Van Niel, on the basis of structural and biochemical comparisons, suggested that two major evolutionary lines appeared to be present. One of them, the *Prokaryota,* contained the bacteria and blue–green algae, and were of relatively simple organization when compared with the second group, the *Eukaryota,* comprising the fungi, algae and protozoa.

Some of the principal structural differences between the two groups are summarized in Table 1.1.

Table 1.1 Some major structural differences between prokaryotic and eukaryotic organisms

STRUCTURE	PROKARYOTE	EUKARYOTE
1. Form of nucleus	Nuclear material present but not bounded by a nuclear membrane.	Always compact nucleus bounded by membrane.
2. Chromosomes	Usually one	Usually more than one
3. Mitotic division	Does not occur	Occurs
4. Mitochondria	Not present	Present
5. Chloroplasts	Not present	Present in photo-autotrophs
6. Vacuoles	Rare	Common
7. Endoplasmic reticulum	Not present	Present
8. Flagella (if present)	One fibril, no membrane	Surrounded by membrane and composed of nine peripheral and two central fibrils
9. Cytoplasmic streaming	Does not occur	Occurs

The viruses are structurally simpler than either the pro- or the eukaryotes. Indeed, whether or not they should be regarded as living organisms is debatable, depending mainly on one's definition of life.

Table 1.2 Units of measurement

1 metre (m) = 1,000 millimetres (mm)
1 mm (10^{-3} m) = 1,000 micrometres (μm)
1 μm (10^{-6} m) = 1,000 millimicrons (mμ)
1 mμ* (10^{-9} m) = 1 nanometre (nm)

* Although the millimicron is not an SI unit it is so widely used in describing the sizes of viruses that its retention here is felt justified.

Table 1.3 Comparative sizes of micro-organisms

ORGANISM		SIZE RANGE
Bacterium	:typical rod	0·5–1·0 × 1·0–10 μm
		(diam.) (length)
Bacterium	:typical sphere	1·0 × 1·0 μm
Fungus	:yeast cell	8–15 × 4–8 μm
Alga	:*Chlamydomonas*	28–32 × 8–12 μm
Virus	:tobacco mosaic virus	300 × 15 mμ

2 BACTERIA

Bacteria were described and drawn by Antonie van Leeuwenhoek as long ago as 1683 but little of significance was known about them until the nineteenth century studies of Louis Pasteur and Robert Koch. Our knowledge of these micro-organisms is now so vast that it embraces a separate branch of biology known as bacteriology.

The following characteristics are typical of bacteria:

1. They are generally unicellular, often existing as colonies. Some species are filamentous.
2. A rigid cell wall usually surrounds the protoplast. In some species the wall is flexible.
3. Reproduction is usually by an asexual process known as fission (Section 2.5.). A few species display budding and others form spores.
4. Some bacteria have photosynthetic pigments and are plant-like in their nutrition. The majority are colourless and feed as saprophytes, parasites or as symbionts.

Included among the parasitic bacteria are those responsible for such human diseases as tuberculosis, leprosy, pneumonia, syphilis, scarlet fever and diphtheria. Fortunately, only a small proportion of the 1,500 known species of bacteria are pathogenic. The remainder play a useful role in the decay of organic matter, such as soil and sewage bacteria, and in the improvement of soil fertility, such as nitrifying and nitrogen fixing bacteria. Man has even harnessed some of the biochemical properties of bacteria in dairying, food processing and other industrial activities.

2.1 TYPES OF BACTERIA

Bacteria are generally placed in the class *Schizomycetes* which is divided into ten orders (Table 2.1).

Table 2.1 Types of bacteria

ORDER	COMMON NAME	CHIEF CHARACTERISTICS	EXAMPLES
Eubacteriales	True bacteria	Unicellular, reproduce by fission, saprophytes, parasites or symbionts. Some species form endospores (survival bodies)	*Azotobacter* (free-living nitrogen fixer) *Rhizobium* (symbiotic nitrogen fixer) *Pasteurella* (cause of plague) *Bacillus* and *Clostridium* (form endospores)
Pseudomonadales	Pigmented sulphur bacteria	Unicellular, reproduce by fission, photosynthetic, chemosynthetic or parasitic. Also includes the stalked iron bacteria	*Chromatium* (purple sulphur bacteria) *Chlorobium* (green sulphur bacteria) *Nitrosomonas* and *Nitrobacter* (nitrifying bacteria) *Desulphovibrio* (sulphate-reducing bacteria)

Actinomycetales	Actinomycetes	Some are branched filaments which produce conidia-like spores. In others the filaments break up into unicells which undergo fission. Includes saprophytes and parasites	*Streptomyces* (soil saprophyte which produces the antibiotic streptomycin) *Mycobacterium* (cause of tuberculosis)
Beggiatoales	Filamentous sulphur bacteria	Chains of colourless cells which display gliding movements. Chemosynthetic, oxidize hydrogen sulphide to sulphur and store the latter as granules in their cells	*Thiothrix, Thiospirillopsis*
Caryophanales	Filamentous sheathed bacteria	Chains of colourless cells containing disc-like nuclear bodies. Found in water containing decomposing organic matter or as intestinal inhabitants	*Caryophanon* (found in cow dung)
Chlamydobacteriales	Filamentous iron bacteria	Chains of colourless cells sheathed in mucilage containing iron oxide. Multiply by means of flagellate swarm spores or by fragmentation of filaments	*Sphaerotilus* (present in water polluted with organic waste)
Hyphomicrobiales	Budding bacteria	Unicells which produce buds at the ends of slender stalks. Some are saprophytic, others photosynthetic	*Hyphomicrobium* (saprophyte) *Rhodomicrobium* (photosynthetic nitrogen fixer)
Mycoplasmatales	Pleuro-pneumonia like organisms (PPLO)	Flexible cell wall, vary in shape (pleomorphic), sometimes filamentous. Filaments break up into elementary bodies which can pass through bacterial filters. Includes saprophytic and parasitic species	*Mycoplasma* (cause of pleuropneumonia in cattle)
Myxobacteriales	Slime bacteria	Unicells having a flexible cell wall. Secrete much slime to form colonies which display creeping movements. Form fruiting bodies bearing microcysts (resting bodies). Saprophytic	*Cytophaga* (cellulose decomposer) *Myxococcus* (digests bacteria)
Spirochaetales	Spirochaetes	Long helical cells with flexible walls. Display squirming movements. Includes saprophytes and parasites	*Treponema* (cause of syphilis) *Borrelia* (cause of relapsing fever) *Leptospira* (cause of leptospiral jaundice)

The first two orders are often described as the true bacteria and the remainder as the higher bacteria. This is an artificial distinction as both groups display prokaryotic characteristics and share many common features. However, as we know so much more about the true bacteria we shall concentrate on these in the remainder of the chapter.

2.2 MORPHOLOGY

Most bacteria are between 0·5 and 1·5 μm in diameter, with a length of several micrometres. Such meagre dimensions give little scope for structural diversity, yet it is the shape of the cell which is first used to classify bacteria (See Fig. 2.1). The cell may be a more or less straight rod, termed a BACILLUS (pl. BACILLI), with a length several times that of its diameter. If the cell is short and comma shaped it is classed as a VIBRIO, if long and cork-screw shaped as a SPIRILLUM (pl. SPIRILLA), and if spherical as a COCCUS (pl. COCCI). During cell division, cocci may divide in one plane and remain attached to form diplococci such as the organism responsible for pneumonia in man, *Diplococcus pneumoniae,* or form chains called streptococci such as the organism responsible for the natural souring of milk, *Streptococcus lactis.* Irregular division of a coccus may produce spheres in grape-like clusters called staphylococci; for example, *Staphylococcus aureus,* which is found on human skin.

Fig 2.1   Common bacterial shapes

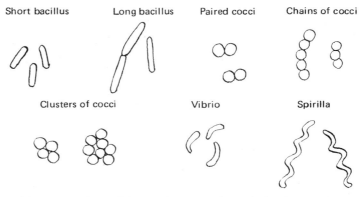

Many species of bacteria may alter their shape either as a result of ageing or because of a sudden environmental shock such as a temperature change. Such abnormalities are called involution forms, and the bacterium is said to exhibit PLEOMORPHY (= several shapes). You can imagine the problems in identification this produces for the microbiologist. As a consequence, all preliminary work in the identification of an unknown species must be concerned with young cultures, growing in a carefully stabilized culture medium.

2.3 CELL STRUCTURE

(a) Cell Wall

Bacteria possess a cell wall, which suggests a closer evolutionary link with the plant kingdom than with the animal kingdom. However, fundamental chemical differences between the cell wall structure of bacteria and higher plants indicate that any relationship is extremely distant.
①The cell wall is a relatively rigid structure which gives shape to the bacterium. It forms about 25% of the dry weight of the cell. Although not essential to life (it can be removed or modified by certain treatments without affecting normal cell activities),②it acts as an outer non-living, but permeable barrier protecting the inner living cell contents③The main chemical component of the wall is a complex polymer, a MUCOPEPTIDE substance④The constituent molecules are two linked amino-sugars: glucosamine and muramic acid (Fig. 2.2). Associated with these amino-sugars are specific amino-acids and frequently proteins and lipids. The whole presents a 3-D, rigid, multi-layered network (Plate 2.1) in marked contrast to the cellulose-

based structure of typical plants. Although the cell wall is difficult to stain it is permeable to solutions of dyes which readily penetrate to the protoplast.

Fig. 2.2 Chemical composition of mucopeptide

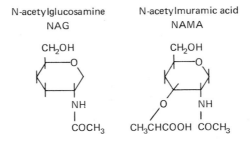

N-acetylglucosamine
NAG

N-acetylmuramic acid
NAMA

Schematic view of bacterial wall organisation

```
NAG       NAG       NAG
  /         /         /
NAMA  =  NAMA      NAMA
  /|        /|        /|
NAG       NAG       NAG
  /         /         /
NAMA      NAMA  =  NAMA
  |         |         |
```

Vertical and double horizontal lines indicate points of attachment of polypeptide chains

Plate 2.1   Isolated cell walls of *Mycobacterium tuberculosis,* × 50,000

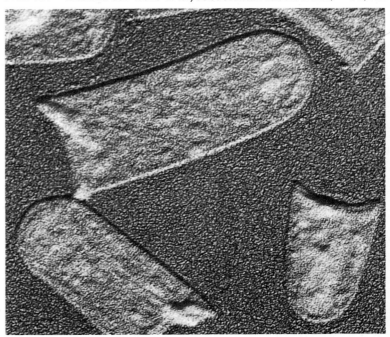

Many bacteria have a CAPSULE of mucilage or slime products deposited around the outside of the cell. The capsules, which may be waste products of metabolism, protect the cell from environmental hazards such as desiccation. The presence of capsules causes the cells to stick together and is responsible for producing the colony form of growth of bacteria on a solid nutrient medium.

(b) Protoplast

The protoplast is the living part of the cell comprising 60–80% of its dry weight. It is always bounded by a PLASMA MEMBRANE, with a structure similar to that of the protein–lipid unit membrane found in eukaryote cells. The membrane is selectively-permeable, controlling the passage of solute molecules in and out of the cell. It may also have the additional function of acting as a centre for cell respiratory activities as do mitochondria in eukaryotes. Invaginations of the plasma membrane, called MESOSOMES, may be concerned with respiration, although these structures may have additional functions such as cell wall synthesis. The cytoplasm appears dense and stains readily with basic dyes such as crystal violet (2.4). The presence of RIBOSOMES, and in photosynthetic species of CHROMATOPHORES containing light-absorbing pigments, has been confirmed by electron microscope studies on ultra thin sections of bacterial cells. Storage products such as volutin, glycogen and starch are often present as INCLUSION BODIES, the amount being dependent on the nutritional level of the organism (see Fig. 2.3 and Plate 2.2).

Fig. 2.3   Structure of a bacterium based on electronmicrographs

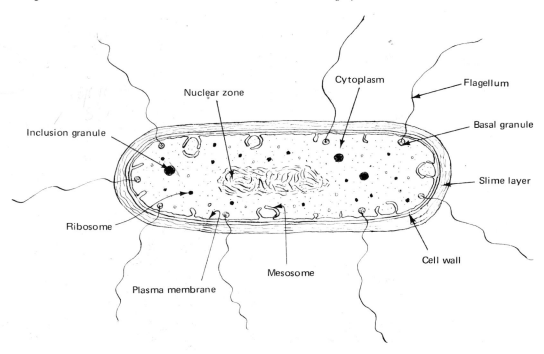

Plate 2.2 Electronmicrograph of a thin section of *Pseudomonas pyocyanea*, x 72,000

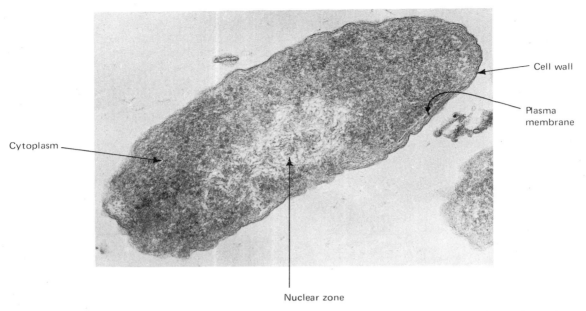

Cell wall

Plasma membrane

Cytoplasm

Nuclear zone

The main difference between eukaryotic and prokaryotic cells centres around the form and organization of the nuclear material. No nucleus comparable to the eukaryotic form has been seen in any prokaryotic cell. The most important differences lie in the lack of a nuclear membrane in prokaryotes and the absence of chromosomes during nuclear division. DNA is the principal component of the nuclear area and can be readily demonstrated using suitable staining techniques. The DNA is in a diffuse zone, the shape and size of which is largely determined by the age and metabolic activity of the cell. The entire NUCLEAR ZONE may represent a single, permanent large molecule of DNA forming a solitary chromosome. In rapidly dividing cells, the nuclear zone may be ahead of vegetative

Plate 2.3 Electronmicrograph of thin section of *P.pyocyanea*, x 72,000

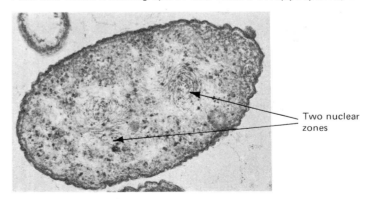

Two nuclear zones

18

division by several generations, with the result that two, four or even eight nuclear areas may occur within a single bacterium for a time (Plate 2.3).

(c) Endospores and Cysts

The ability to form ENDOSPORES which are resistant to high temperatures and toxic chemicals is characteristic of certain species of rod-shaped bacteria of the genera *Bacillus* and *Clostridium* (Fig. 2.4). Such spores are not reproductive units like the spores of fungi or higher plants since one cell produces only one endospore which germinates to produce one new cell. The size, shape and position of the spore in the cell are useful diagnostic features.

Fig. 2.4   Positions of endospores in bacteria

Bacillus sp.   *Clostridium* sp.

Terminal Central

Some spores have a very long life. Anthrax spores in the soil can live for over 50 years. The longevity is due to an extremely resistant spore coat, which may be of several layers. In addition, the abnormally low water content of the spore cytoplasm (15% of vegetative cytoplasm) reduces the enzyme activity of the cell. The bacterial endospore literally represents suspended animation, and can germinate when conditions for growth are ideal (Fig. 2.5).

Fig. 2.5   Structure of bacterial endospore

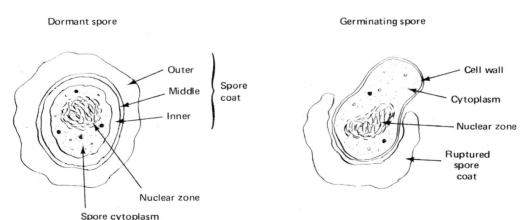

Dormant spore

Outer
Middle } Spore coat
Inner

Nuclear zone

Spore cytoplasm

Germinating spore

Cell wall

Cytoplasm

Nuclear zone

Ruptured spore coat

Some soil-dwelling bacteria such as *Azotobacter*, unable to form an endospore, round off and produce a CYST. The cell shortens and the cell wall becomes thickened. Cysts have limited powers of resistance to unfavourable conditions. On germinating they give rise to new vegetative cells.

(d) Flagella and Motility

Many species of bacteria can move rapidly in a liquid medium by means of FLAGELLA (sing. FLAGELLUM). A flagellum is a thread-like, unbranched structure about 20–30 μm long. Flagella require special treatment if they are to be observed with a light microscope. A cell may have one or many flagella, which may be distributed over the whole surface (peritrich form), or present at one or both ends of the cell (polar form) (Fig. 2.6). Both position and number of flagella are important identification characters.

Examination of flagella under the electron microscope (Plate 2.4) shows them to lack the characteristic 9:2 microtubule organization of the eukaryote flagella. They are composed solely of a unique type of protein termed flagellin. Like the protein myosin of muscle fibre, flagellin has the special property of contraction. Each flagellum is anchored inside the protoplast by a small basal granule. How the flagellum moves is not yet known. Fimbriae are shorter structures similar to flagella. They are unable to make the cell move, but attach some species to their substrate (Plate 2.5).

Plate 2.4   Electronmicrograph of *Pseudomonas* sp. showing flagella, × 50,000

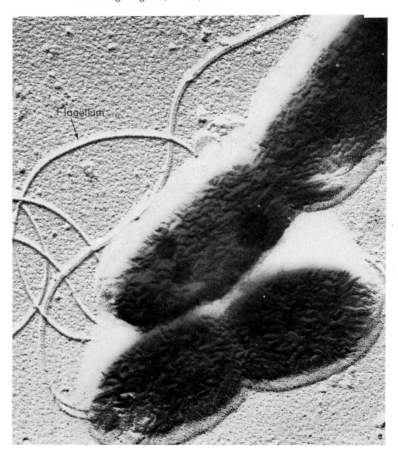

Flagellum

20

Fig. 2.6 Arrangement of bacterial flagella

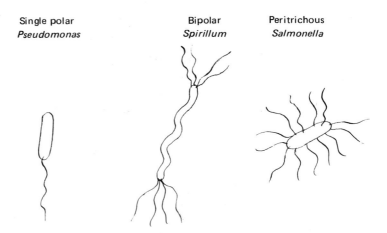

Single polar
Pseudomonas

Bipolar
Spirillum

Peritrichous
Salmonella

Plate 2.5   Electronmicrograph of a metal shadowed preparation of cells of *Escherichia coli* surrounded by fimbriae, × 10,000

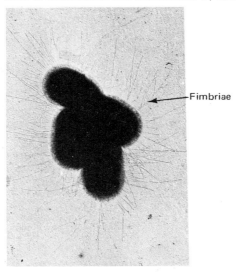

—Fimbriae

2.4 STAINING PROPERTIES

The treatment of dried, fixed bacterial cells with aniline dyes shows important details of the cell structure. The bacteriologists of the nineteenth century perfected a variety of staining procedures which are still used to demonstrate specific morphological characters. Gram's method is the most widely used of these procedures. It was developed by Christian Gram in 1884 and has undergone minor modifications since that time. It enables one to subdivide bacteria into two groups: Gram positive and Gram negative. Details of the Gram staining procedure are given in section 10.5.1. The complete explanation for the Gram reaction is not yet known. The ease with which the crystal violet iodine complex is removed from Gram negative cells suggests fundamental differences in their wall structure when compared with positive forms. Certainly the mucopeptide of the walls of Gram negative bacteria contains up to 20% lipids, which are hardly present in the walls of Gram positive organisms.

Old cultures of Gram positive bacteria often give an indeterminate or even negative reaction. This again points to the necessity of using young cultures for experimental purposes. Although the division of the bacterial world into two groups on the basis of the Gram reaction is purely arbitrary, it does coincide with other characteristics. All sporing bacteria, for example, are Gram positive, all bacteria with polar flagella are Gram negative.

2.5 REPRODUCTION

Under favourable conditions cells grow quickly and soon reach their maximum size. For certain bacteria growing under ideal conditions, the generation time, the time taken for one cell to enlarge and divide into two, may be no longer than twenty minutes. For other species it can be as long as fifteen to twenty hours. The process of division is known as BINARY FISSION (Fig. 2.7). First the nuclear area becomes constricted and a cross wall begins to grow inwards. The continued growth of the cross wall eventually cleaves the protoplast so that two daughter cells are formed (Plate 2.6).

Fig. 2.7   Binary fission

Bacillus

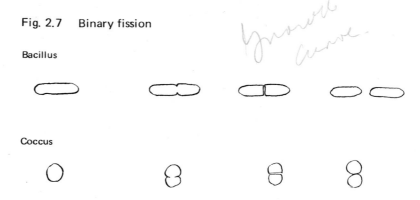

Coccus

This multiplication process is by far the most common means of reproduction displayed by bacteria. Since the DNA replicates prior to fission, the progeny will have the same genotype as the parent unless a mutation has occurred. However, for certain bacteria such as *Escherichia coli,* genes can be transferred from one cell to another by a process of CONJUGATION during which cytoplasmic connections are made between mating cells. Genes pass from a donor cell to a recipient cell and following separation the latter displays new genetic properties.

Another process leading to the formation of new bacterial genotypes is that of TRANSFORMATION whereby DNA released into the medium by one cell may be absorbed by another bringing about a genetic change.

Viruses may also be concerned in altering the genome of a bacterial cell by means of TRANSDUCTION (see Chapter 5).

Plate 2.6 Electronmicrographs of thin sections of
Micrococcus lysodeikticus showing binary fission, x 47,000

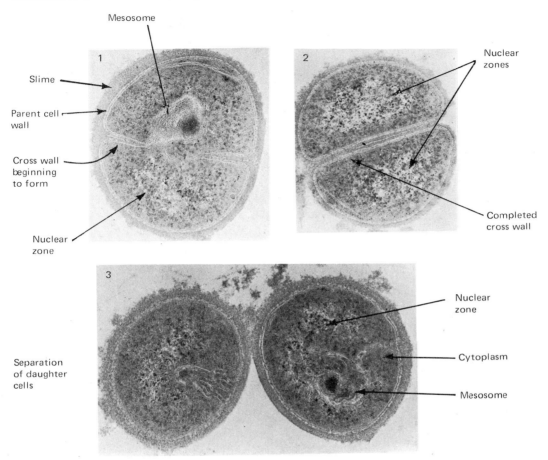

Mesosome

Slime

Parent cell
wall

Cross wall
beginning
to form

Nuclear
zone

Nuclear
zones

Completed
cross wall

Separation
of daughter
cells

Nuclear
zone

Cytoplasm

Mesosome

2.6 NUTRITION

All living organisms need at least two basic ingredients from
the environment: the raw materials necessary for the building
of cell structures, and a source of energy. Provided these
requirements are satisfied the organism is able to grow and
ultimately increase in numbers.

Most bacteria, like all animals, are HETEROTROPHS. Their source
of energy is previously synthesized foodstuffs, especially carbo-
hydrates. Amongst the bacteria are non-fastidious heterotrophs
such as *Bacillus spp.* which can grow on simple sugars and
mineral salts alone. Fastidious heterotrophs, for example
Lactobacillus spp., require the addition of complex nutrients
such as vitamins and amino-acids before growth occurs.

The wide variations in nutritional requirements, reflecting
biochemical (enzyme) differences, may be used to distinguish
between different species. Other bacteria are AUTOTROPHS and
these synthesize complex growth requirements from inorganic
nutrients of which carbon dioxide is the principal source of

carbon. Some of these forms obtain their energy from sunlight (PHOTOAUTOTROPHS), others produce chemical energy (CHEMO-AUTOTROPHS) in order to fix carbon dioxide.

(a) Photoautotrophic bacteria

Bacterial photosynthesis follows in outline the photosynthetic pathways of the higher plants and green algae, although differences exist between the photosynthetic processes of prokaryotes and eukaryotes. The differences are listed in Table 2.2.

Table 2.2 Distinction between photosynthesis in bacteria and eukaryotes

	PHOTOSYNTHETIC BACTERIA	EUKARYOTES
Nature of pigment	Bacteriochlorophyll	Chlorophyll
Colour of pigment	Green and/or purple	Green
Electron source	Hydrogen sulphide and/or organic compounds	Water
Relationship to oxygen	Anaerobic	Aerobic
Maximum wavelength of absorption	Infra-red 730–900 nm	Red 680 nm

The ecological significance of members of this group is difficult to assess. They probably contribute little to the overall food production in their ecosystem since their biomass remains small in comparison with higher plants. Neither do they produce oxygen in photosynthesis.

On the basis of pigment colour and substrate used, the photoautotrophic bacteria fall into three groups:

1. Green sulphur bacteria, e.g. *Chlorobium*.
2. Purple sulphur bacteria, e.g. *Chromatium*.
3. Purple, non-sulphur bacteria, e.g. *Rhodospirillum*.

Bacteria of groups 1 and 2 use a common electron source of hydrogen sulphide or other sulphur compounds. They differ from each other in the nature of pigments accessory to chlorophyll, for example, carotenoids which are present in the cytoplasm. The photosynthetic reactions may be represented by the following equations:

$$\text{bacteria: } CO_2 + 2H_2S \xrightarrow[\text{bacterio-chlorophyll}]{\text{light}} CH_2O + H_2O + 2S$$

$$\text{green plants: } CO_2 + 2H_2O \xrightarrow[\text{chlorophyll}]{\text{light}} CH_2O + H_2O + O_2$$

Members of the third group are not strict autotrophs since they depend on energy-rich organic substrates such as alcohols to supply the hydrogen required for the fixation process and can grow as heterotrophs in the dark.

(b) Chemoautotrophic bacteria

These use carbon dioxide and other simple inorganic constituents to synthesize organic materials. The energy necessary for the syntheses is derived from oxidation or reduction reactions. Certain members of this group play an important ecological role in the recycling of essential mineral nutrients such as nitrogen, sulphur and iron. Among the bacteria involved in the circulation of nitrogen are *Nitrosomonas* and *Nitro-*

sococcus, both of which oxidize ammonium salts to nitrites under aerobic conditions:

$$2NH_4^+ + 3O_2 \rightarrow 2NO_2^- + 4H^+ + 2H_2O + energy$$

The bacterium *Nitrobacter*, under similar conditions, brings about further oxidation of the nitrite to nitrate, the major form of nitrogen taken in by the roots of higher plants.

$$2NO_2^- + O_2 \rightarrow 2NO_3^- + energy$$

Thiobacillus, another chemoautotrophic bacterium, obtains its energy supplies by the oxidation of soil sulphur to sulphate:

$$2S + 3O_2 + 2H_2O \rightarrow 2H_2SO_4 + energy$$

Sulphur oxidation also plays a very important part in the maintenance of soil fertility since the oxidized state as sulphate is the most readily available form of sulphur for plants and other micro-organisms.

The link between heterotrophic and autotrophic bacteria lies in the molecule adenosine triphosphate (ATP). The formation of this substance is common to all forms of life, bacterium and man alike. The many differing metabolisms exhibited amongst bacteria have one basic function, to provide the organism with a supply of energy in the form of ATP. This may be achieved in heterotrophs by breaking down a simple sugar; in photoautotrophs by harnessing the energy from sunlight via an intermediate pigment system; or in chemoautotrophs from the oxidation of inorganic components.

2.7 RESPIRATION

For most plants and animals oxygen is necessary for life. Their basic energy-producing reactions (aerobic respiration) will continue to operate only in the presence of oxygen and if deprived of it they die. Such organisms are OBLIGATE AEROBES. Many bacteria, especially those living in soil, are obligate aerobes. However certain bacteria, called FACULTATIVE AEROBES, appear to be indifferent to the presence of oxygen in their environment. In contrast, some bacteria are actually inhibited by oxygen, which causes enzyme inactivation. They may even die on exposure to air. These are the strict or OBLIGATE ANAEROBES, such as *Clostridium*, which lives in the soil. Further details of microbial respiration are discussed in Chapter 7.

Different environments contain different amounts of oxygen. In well-drained, loamy agricultural soil, oxygen is freely available and chemoautotrophs such as *Nitrosomonas* and *Thiobacillus* can effect their oxidative reactions, produce energy and thus thrive. For bacteria living in the mammalian gut however the amount of oxygen will be low and facultative aerobes such as *Escherichia coli* are dominant. The mud at the bottom of a pond or lake, rich in organic pollutants, will contain little or no free oxygen, and obligate anaerobes such as the photosynthetic bacterium *Chromatium* will dominate the bacterial population.

Even within the same habitat, the amount of oxygen may fluctuate and enable different species to be active at different times. A well-drained soil, for example, following heavy rain, may become temporarily anaerobic and permit the growth of species such as the denitrifier *Pseudomonas denitrificans* which reduces nitrate to ammonia.

2.8 BACTERIA AND NUTRIENT CYCLES

The role of heterotrophic bacteria in changing organic compounds into simpler inorganic substances, and of autotrophic bacteria in converting one inorganic form of an element into another has already been briefly mentioned (see Section 2.6). The consequence of such activities is that microorganisms prevent the accumulation of organic matter and ensure that supplies of essential nutrients such as nitrates, sulphates and phosphates are constantly available to green plants.

(a) The nitrogen cycle

Fig. 2.8 The nitrogen cycle

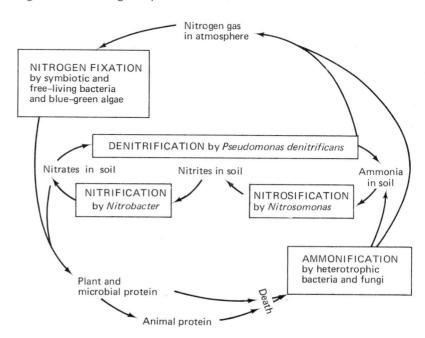

Life is impossible without nitrogen but the form in which this element is used by living organisms differs considerably. In its most common state as nitrogen gas it is biologically inert except to a few nitrogen-fixing micro-organisms which live mainly in soil. Under natural conditions many organic nitrogenous compounds such as proteins, amino-acids, amides and nucleic acids are returned to the soil in dead organic matter and faeces. Such complex materials provide nutrients for many heterotrophic soil microbes, which by their activities produce ammonia and nitrate. These inorganic nitrogenous substances are then available for higher plants to absorb and utilize for protein and nucleic acid synthesis. The microbially controlled processes occur in a number of stages (Fig. 2.8) and such is the rate of the changes that the annual global turnover of soil nitrogen is of the order of 10^9 and 10^{10} tonnes per annum.

26

(*i*) *Ammonification.* Aerobic bacteria such as *Pseudomonas* sp. and fungi such as *Penicillium* and *Mucor* are examples of the many species of heterotrophic soil organisms which decay organic matter with the release of ammonia. Secretions of peptidase enzymes hydrolyse proteins to amino-acids, some of which are absorbed and used to synthesize microbial protein. Unwanted amino acids are deaminated and ammonia is formed by a comparable process to that occurring in the mammalian liver.

$$\underset{\underset{COOH}{|}}{\overset{\overset{R}{|}}{CHNH_2}} + O \rightarrow \underset{\underset{COOH}{|}}{\overset{\overset{R}{|}}{CO}} + NH_3$$

The familiar smell in the vicinity of manure heaps indicates that some ammonia escapes into the atmosphere when organic matter is decayed. However, much ammonia enters the soil and is absorbed by plant roots or is converted to nitrate by nitrifying bacteria.

(*ii*) *Nitrification.* In aerobic conditions such as would prevail in a well-drained soil, ammonia is oxidized to nitrate by two highly specialized groups of chemo-autotrophic bacteria, the NITROSIFIERS and the NITRIFIERS. Nitrosification is the oxidation of ammonia to nitrite:

$$2NH_4^+ + 3O_2 \rightarrow 2NO_2^- + 4H^+ + 2H_2O + energy$$

The principal bacterium concerned with this conversion is *Nitrosomonas.* Nitrite is toxic to higher plants but it is oxidized as quickly as it is formed into nitrate by bacteria such as *Nitrobacter*:

$$2NO_2^- + O_2 \rightarrow 2NO_3^- + energy$$

Although nitrification does not actually increase the nitrogen content of soil it results in the production of nitrate which is more readily available to higher plants as a nitrogen source.

(*iii*) *Denitrification.* This occurs in waterlogged soils and leads to a loss of biologically important forms of soil nitrogen. In such conditions anaerobic bacteria, for example, *Pseudomonas denitrificans,* use nitrate as an electron acceptor during respiration. The consequence of this is the reduction of nitrate to ammonia and gaseous nitrogen with an attendant fall in soil fertility.

(*iv*) *Nitrogen fixation.* This involves the biochemical reduction of gaseous nitrogen to form ammonia:

$$N_2 + 3H_2 \rightarrow 2NH_3$$

The micro-organisms involved are principally two groups of heterotrophic bacteria. SYMBIOTIC nitrogen fixers belong to the genus *Rhizobium* which infects the root hairs of higher plants, mainly legumes such as peas, beans and clover. The roots are then stimulated to form swellings or nodules inside which the bacteria multiply and fix gaseous nitrogen. The association is one of mutual benefit; the leguminous plant has a reduced form of nitrogen immediately available for protein and nucleic acid synthesis whilst the bacterium obtains sugars from the plant root. *Rhizobium* cannot survive in soil for any length of time without the host plant. For this reason legume seed is usually inoculated with the bacterium before it is sown. Such is the effectiveness of the association that no artificial fertilizers are needed to maintain soil fertility in clover and grass pastures in New Zealand.

FREE-LIVING nitrogen-fixing bacteria include the aerobic *Azotobacter* and the anaerobic *Clostridium* both of which live in soil humus. In soils which are favourable for their growth they can fix as much nitrogen as their symbiotic counterparts. Blue–green algae such as *Nostoc* are also important nitrogen fixers (Chapter 3).

Although the activities of soil microbes are generally beneficial to higher plants there are occasions when they can IMMOBILIZE soil nitrogen. When carbon-rich crop residues such as straw are ploughed into soil, bacteria and fungi multiply very quickly and compete with plant roots for ammonia and nitrate. This may result in a decreased yield in a subsequent crop unless additional fertilizers are used. Immobilization does not cause a lowering of the total soil nitrogen level, it simply results in a temporary decrease of available nitrogen. When the micro-organisms die their organic nitrogenous materials are decayed to release ammonia.

(b) The sulphur cycle

Fig. 2.9 The sulphur cycle

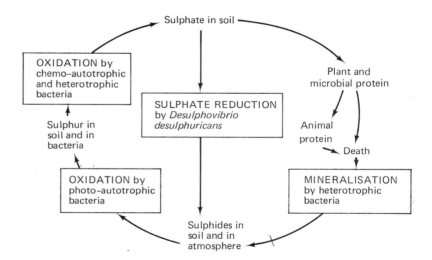

Sulphur is present in some amino-acids such as methionine and cystein which form an integral part of many proteins. Like nitrogen, therefore, it is essential for life. Sulphate acts as the major source of sulphur to higher plants, yet dead plant and animal material entering soil contains organic sulphur in proteins and amino-acids. The transformation of organic sulphur to inorganic sulphate is dependent on micro-organisms and in many ways is similar to the recycling of nitrogen (Fig. 2.9). The principal stages are:

(*i*) *Mineralization*. The breakdown of organic sulphur compounds by heterotrophic bacteria results in the formation of hydrogen sulphide gas and sulphur, a process comparable to ammonification. Some of the gas escapes into the atmosphere but the remainder can be oxidized to sulphur by photo-autotrophic bacteria such as *Chlorobium* and *Chromatium*:

$$\text{Sun energy} + 2H_2S + CO_2 \rightarrow CH_2O + 2S + H_2O$$

(*ii*) *Sulphur oxidation*. This is comparable to nitrification and results in the production of sulphate from hydrogen sulphide and sulphur. The organisms responsible include heterotrophic bacteria and fungi and also chemoautotrophs belonging to the genera *Thiobacillus*, *Thiothrix* and *Beggiatoa*:

$$2S + 2H_2O + 3O_2 \rightarrow 2H_2SO_4 + \text{energy}$$

The sulphate is now available for root absorption by higher plants.

(*iii*) *Sulphate reduction*. This is analogous to denitrification and is brought about by anaerobic bacteria such as *Desulphovibrio desulphuricans*:

$$4H_2 + H_2SO_4 \rightarrow 4H_2O + H_2S$$

Although many heterotrophic soil microbes prefer organic forms of sulphur they can also use sulphate. Thus IMMOBILIZATION of sulphur may follow the enrichment of soil with carbon-rich crop residues.

3 THE BLUE–GREEN ALGAE

Blue–green algae are placed in the class *Cyanophyta* (*Myxophyceae*). They are immediately distinguished from other algae because of their structural simplicity. Most of the *Cyanophyta* are blue–green in colour, others are red, yellow, brown or even colourless. Colour is not only dependent on the nature of the pigments present but also on the age, physiological state and environmental conditions under which the organism is living. Three pigments common to all photosynthetic plants are present: chlorophyll, carotene and xanthophyll. In addition, two pigments unique to blue–greens, blue PHYCOCYANIN and red PHYCOERYTHRIN, are responsible for the wide range of colour.

There are many similarities between blue–greens and bacteria. These include:

1. Nuclear material lacking a membrane.
2. No typical mitotic stages during cell division.
3. Absence of mitochondria and other internal membraneous structures.
4. No apparent sexual stages in the life-cycle.

They are also physiologically similar to some bacteria in that some blue–greens can live anaerobically in mud at the bottom of ponds and rivers and exchange a photosynthetic existence for a heterotrophic one. Others can fix atmospheric nitrogen and help to increase soil fertility as several groups of bacteria do. The blue–greens then, are thought to be related to the bacteria from an evolutionary standpoint, and both groups are regarded as being at a more primitive level of organization than all other organisms. Fossil evidence from rocks suggests that the earliest forms of life on earth may well have been like the blue–green algae in existence today.

3.1 MORPHOLOGY AND STRUCTURE

There are about 2,000 species of blue–green algae growing either as unicells or multicells (Fig. 3.1). The unicells are either rods or spheres whilst the multicells are hair-like filaments.

The presence of mucilage, often produced as a sheath outside the cell wall, causes the cells to adhere as a slimy or jelly-like growth on soil or on the surface of water. The thin cell wall is composed of a mucopeptide substance of similar composition and arrangement to that found in bacterial walls. Little of the protoplast can be made out with the light microscope other than a central region, the NUCLEOPLASM containing nuclear material without a membrane, and an outer CHROMATO-PLASM containing dispersed pigments. The electron microscope shows that this simple organization is in fact complex, with a series of lamellae or membranes which contain the photosynthetic pigments, extending through the cytoplasm (Plate 3.1). Most species have cell inclusions usually of reserve food

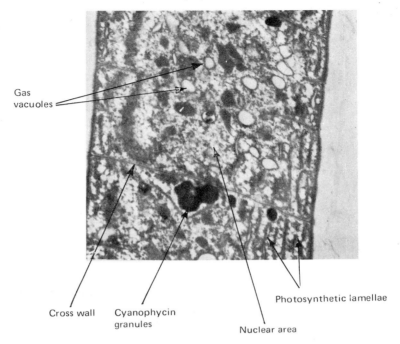

Gas
vacuoles

Cross wall

Cyanophycin
granules

Nuclear area

Photosynthetic lamellae

Fig. 3.1 Range of form of some blue-green algae

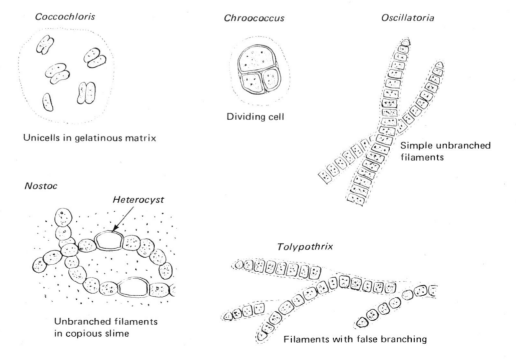

Coccochloris

Unicells in gelatinous matrix

Chroococcus

Dividing cell

Oscillatoria

Simple unbranched
filaments

Nostoc

Heterocyst

Unbranched filaments
in copious slime

Tolypothrix

Filaments with false branching

31

materials such as cyanophycin granules. In aquatic forms, gas vacuoles give buoyancy to the cell enabling it to float. Many filamentous blue—green algae form special thick-walled cells known as HETEROCYSTS which at maturity have transparent protoplasts (Fig. 3.1).

3.2 REPRODUCTION

Reproduction in the blue—green algae is by an asexual process comparable to fission in bacteria. Growth of the daughter cells in filamentous species brings about an increase in length, followed by the breakage of the filament. The term HORMOGONIUM is used to describe a fragment of a filament containing a small number of cells.

Breakage of the filament with formation of hormogonia often takes place at the points where heterocysts occur. It has been suggested that the heterocyst may be a primitive form of reproductive cell still retained by this group of plants. On the other hand, heterocysts may well be a physiological specialization of the vegetative cell. They lack the pigments necessary for photosynthesis, but contain enzyme systems capable of nitrogen fixation. Some filamentous forms such as *Anabaena* produce enlarged, thick-walled cells (AKINETES) which germinate to form new filaments.

3.3 NITROGEN FIXATION AND ECOLOGY

Many filamentous, heterocyst-containing blue—green algae can fix atmospheric nitrogen. This, coupled with their role as primary producers of energy via photosynthesis, make their contribution to soil or pond fertility of considerable importance. Non-symbiotic nitrogen fixation by species such as *Nostoc* and *Tolypothrix* is believed to maintain the fertility of many tropical soils; as much as 625 kg of nitrogen gas may be fixed per km^2 annually in paddy fields. The addition of blue—green algae as fertilizers to barren Indian soils of poor texture increased the soils' nitrogen and humus contents, and improved their water-holding capacities so much that they could support vigorous crop growths. Asian paddy fields contain extensive natural surface blooms of blue—green algae which are believed to be responsible for the continued successful cultivation of rice without added nitrogenous fertilizers.

In distribution, the *Cyanophyta* occur in every habitat known to support life, ranging from hot sulphur springs at temperatures up to 70°C to arctic waters permanently below freezing point. Their growth under extreme conditions enables them to be the first inhabitants of bare rock and soil. This ability to act as pioneer organisms was recently demonstrated by their dramatic colonization of the rock of Surtsey, a new volcanic island which appeared off Iceland in 1963. Because they can photosynthesize and fix atmospheric nitrogen the blue—green algae have the simplest possible nutritional requirements. Given light, minerals, nitrogen, carbon dioxide and water they can thrive in conditions impossible for most other forms of life. Several species of blue—greens are involved in symbiotic associations with a variety of other plants such as lichens (Chapter 6).

Aggregations of blue—green algae forming scums or blooms in aquatic situations sometimes present ecological problems (Chapter 7). The Red Sea is said to have obtained its name from the reddish-tinged blooms of *Trichodesmium* which appear in it from time to time.

4 PROTOZOA

The *Protozoa* is an extremely diverse and varied group of eukaryotic micro-organisms ranging from the relatively simple *Amoeba* to the highly specialized *Paramecium*. Because they move and do not photosynthesize most protozoa are considered to be members of the animal kingdom; yet this group contains some species such as the *Euglena* where distinction between the plant and animal kingdoms breaks down.

With over 30,000 named species the protozoa are widely distributed on land and in water. They are often closely associated with other animals including man. Their occurrence may be casual and harmless, for example many gut protozoa, or pathogenic such as the malarial parasite.

4.1 TYPES OF PROTOZOA

The phylum is usually divided into four classes (Table 4.1).

Table 4.1 Classification of the Protozoa

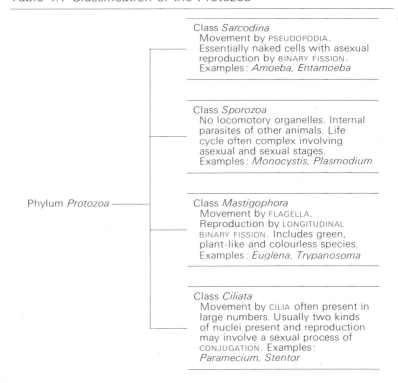

Phylum *Protozoa*

Class *Sarcodina*
Movement by PSEUDOPODIA. Essentially naked cells with asexual reproduction by BINARY FISSION. Examples: *Amoeba, Entamoeba*

Class *Sporozoa*
No locomotory organelles. Internal parasites of other animals. Life cycle often complex involving asexual and sexual stages. Examples: *Monocystis, Plasmodium*

Class *Mastigophora*
Movement by FLAGELLA. Reproduction by LONGITUDINAL BINARY FISSION. Includes green, plant-like and colourless species. Examples: *Euglena, Trypanosoma*

Class *Ciliata*
Movement by CILIA often present in large numbers. Usually two kinds of nuclei present and reproduction may involve a sexual process of CONJUGATION. Examples: *Paramecium, Stentor*

4.2 CLASS *SARCODINA* (*RHIZOPODA*)

The *Sarcodina* are the least specialized of the protozoa. The majority are free-living organisms but there are a few parasitic species such as *Entamoeba histolytica,* the cause of amoebic dysentery in man. Many of the free-living rhizopods which form part of zooplankton are protected by a shell or supported by a siliceous skeleton but the amoebae are little more than naked protoplasts.

(a) Structure

Plate 4.1   Phase contrast photomicrograph of *Amoeba proteus* × 180

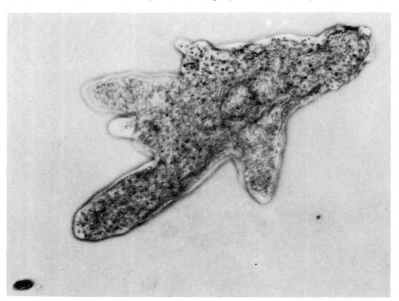

Fig. 4.1   Structure of *Amoeba proteus*

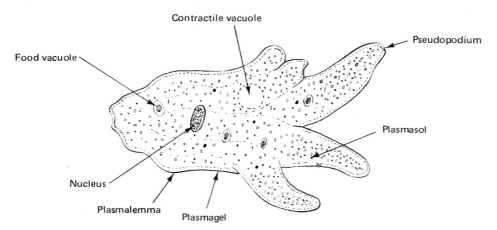

Under the light microscope the large freshwater *Amoeba proteus* looks like a colourless blob of jelly which is constantly changing its shape by a flowing or rolling action (Plate 4.1). A thin elastic membrane, the PLASMALEMMA, encloses the CYTOPLASM which has an outer thin transparent layer or PLASMAGEL surrounding the bulky, granular PLASMASOL. In the plasmasol are a disc-shaped NUCLEUS, VACUOLES containing particles of food, and a single large CONTRACTILE VACUOLE which has an osmo-regulatory function. INCLUSIONS such as oil globules and crystals of varied chemical composition are present in amounts determined by the age and activities of the cell (Fig. 4.1).

(b) Movement

The animal moves in an irregular flowing manner termed AMOEBOID MOVEMENT, which results from the formation of PSEUDOPODIA (sing. PSEUDOPODIUM) which are temporary finger-like extensions of the cytoplasm. This type of movement is common in certain tissues of multicellular animals; for example, white blood cells of vertebrates. The mechanism by which this movement is brought about has been the subject of much controversy and several theories have been proposed to explain it. During movement, the cell has a definite polarity; there is an anterior end from which one or more pseudopods are formed, and a posterior end where they are lacking. Locomotion is thought to depend on changes in viscosity of the cytoplasm which is effected by sol–gel transformations. The plasmagel softens at a point where a pseudopodium is to be formed and plasmasol flows towards this point of weakness causing it to bulge outwards. Near the tip of the pseudopod sol is converted to gel, whilst at the posterior end of the body gel is changed to sol which flows forwards (Fig. 4.2). It has been suggested that the sol–gel changes are brought about by alternate relaxation and contraction of large protein molecules. In the gel the molecules extend to form crosslinkages with other molecules creating a rigid framework; in the sol they become coiled and compact with few crosslinkages thus forming a semi-fluid matrix. This change in state is an energy-requiring process and considerable ATP activity has been demonstrated in the tail region.

Fig. 4.2   Amoeboid movement

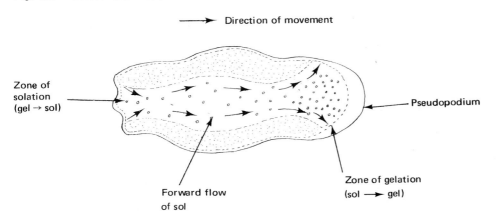

Direction of movement

Zone of
solation
(gel → sol)

Pseudopodium

Forward flow
of sol

Zone of gelation
(sol ⟶ gel)

Other workers maintain that the site of contraction is at the anterior end of the cell. Here, protein molecules, initially in an extended state in the plasmasol, become condensed into plasmagel. This results in a force which sucks the plasmasol to the anterior end. This idea is generally referred to as the fountain-zone theory. Another suggestion is that inner layers of gel molecules push the sol molecules along by a kind of ratchet mechanism.

Whichever explanation best fits the mechanism of locomotion in *Amoeba*, it is certain that contact with a solid surface such as the bottom of a pond, or floating leaf debris, is essential to bring about movement. Away from such surfaces, *Amoeba* does not exhibit the formation of pseudopodia.

(c) Nutrition

Amoeba feeds on a varied diet of pond micro-organisms. Small algae and especially live ciliates and flagellates serve as the major food source. The proximity of food stimulates the formation of a pseudopodium which, by membrane invagination, forms a cup-like depression around the prey to trap it. The food-cup finally closes and the prey is taken into the cell with some water as a food vacuole or phagosome. This method of feeding is termed PHAGOCYTOSIS (Fig. 4.3). The vacuoles are moved around the cytoplasm and enzyme activity brings about digestion of the food. Initially the food vacuoles display an acid reaction which gradually becomes alkaline as the prey is broken down. Eventually the soluble food products are absorbed into the cytoplasm and undigested residues are passed to the cell surface to be left behind in the water.

Fig. 4.3   Phagocytosis

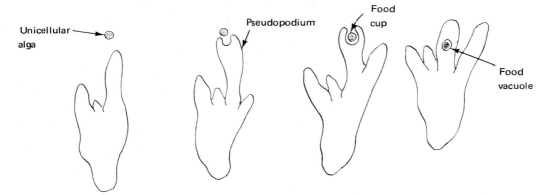

Amoeba may also take in water droplets containing soluble nutrients during PINOCYTOSIS or cell drinking, another characteristic activity of the cell membrane.

(d) Reproduction

On reaching its optimum size the *Amoeba* becomes spherical and mitosis occurs (Fig. 4.4). This process of BINARY FISSION may occur every two or three days in cultures maintained under favourable conditions. Reports of *Amoeba* producing cysts or spores have now been disproved but in the related genus *Entamoeba* resistant cysts are formed. They are passed in the faeces of an infected person and are responsible for the spread of amoebic dysentery.

Fig. 4.4   Binary fission in *Amoeba*

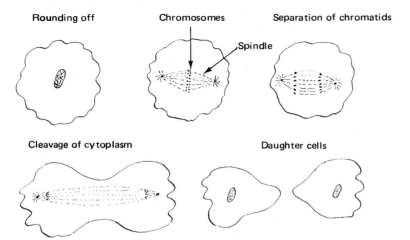

Rounding off   **Chromosomes**   **Separation of chromatids**

Spindle

Cleavage of cytoplasm   **Daughter cells**

4.3 CLASS MASTIGOPHORA (FLAGELLATA)

Colourless flagellates such as *Trypanosoma,* the cause of sleeping sickness, are placed in the sub-class *Zoomastigina*. *Euglena* is a member of the chlorophyll-bearing group of flagellates, the *Phytomastigina*. These organisms photosynthesize but if kept in darkness they may become heterotrophic. The ability of organisms such as *Euglena* to feed both auto- and heterotrophically suggests that the division between the plant and animal kingdoms is not rigid at this level. Thus the green flagellates can also be regarded as algae and classed in the *Euglenophyta*.

(a) Structure

Euglena is a common member of the phytoplankton of fresh and salt waters and is also found on damp soil. The body is slender and elongate (Fig. 4.5) with a blunt anterior end and pointed posterior. The shape is maintained by a flexible PELLICLE forming an outer covering membrane, which winds spirally around the body. The anterior end has a flask-shaped depression made up of a funnel-like mouth or CYTOSTOME and a short canal leading into a RESERVOIR. The latter leads into a CONTRACTILE VACUOLE at its innermost surface. Several smaller accessory vacuoles complete the somewhat complex osmoregulatory apparatus. A STIGMA or eye spot containing the light-sensitive pigment carotene lies to one side of the gullet.

The FLAGELLUM is in fact two flagella of unequal length which arise from the base of the reservoir. One flagellum extends through the cytostome to form the whip-like locomotory structure, readily observed under the light microscope. The second flagellum is much shorter and remains inside the reservoir lying close to the base of the long flagellum. Both flagella have the typical 9+2 internal organization of filaments in common with eukaryotic cells, and arise from basal granules or BLEPHAROPLASTS (Plate 4.2). The cytoplasm contains numerous mitochondria, a nucleus, food granules of a starch-like polysaccharide (paramylum) and chloroplasts. The shape and number of the chloroplasts vary with species. In *E. viridis*, they form a stellate structure radiating from the centre of the cell. In *E. gracilis,* they are large flattened plates extending throughout the protoplast. Each chloroplast has a PYRENOID which stores paramylum.

Fig. 4.5   Structure of *Euglena viridis*

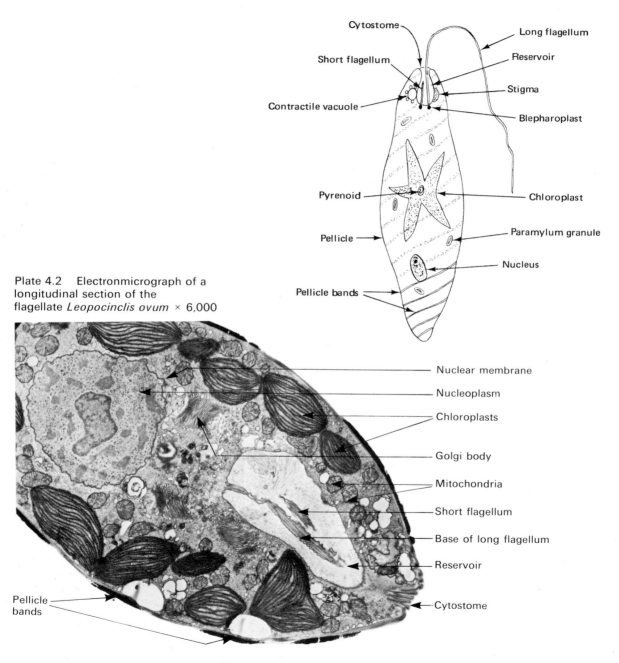

Plate 4.2   Electronmicrograph of a longitudinal section of the flagellate *Leopocinclis ovum* × 6,000

(b) Movement

Waves of contraction spread from the base of the flagellum to its tip. This creates a force which pulls the cell through the water. The waves proceed along the flagellum in a spiral manner so that, as *Euglena* moves forward, it traces out a helical path. At the same time the cell rotates about its own axis (Fig. 4.6). The energy for the contractions is derived from ATP which is present in the flagellum. Use of the flagellum results in rapid movement of the organism.

Fig. 4.6   Flagellate movement of *Euglena*

Movement of a very different nature is also possible in *Euglena* as a result of alternate contractions and relaxations of the flexible pellicle. Strips of pellicle material are wound around the cell body below the plasma membrane. Articulation between the interfaces (PELLICLE BANDS) of the strips allows a gradual change in body shape to occur and slow, creeping euglenoid movements occur (Fig. 4.7).

Fig. 4.7   Euglenoid movement

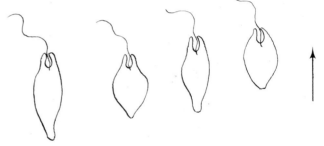

Euglena normally swims towards a source of light with its long axis parallel to the light rays. The PHOTOTACTIC response is controlled by a PHOTORECEPTOR at the base of the longer flagellum working in conjunction with the stigma. If the stigma temporarily shields the photoreceptor from light, as for example when light rays are at right angles to the body, *Euglena* changes direction and swims parallel to the light once more.

(c) Nutrition

When light is present *Euglena* photosynthesizes and is therefore photoautotrophic. It does, however, depend on certain growth factors such as vitamins B_1 and B_{12} being available, without which growth does not occur. In the continued absence of light, breakdown of the photosynthetic pigments occurs and nutrition then becomes heterotrophic if organic materials are available in a soluble state. On return to sunlight such bleached cells recover their greenness and photosynthetic ability.

(d) Reproduction

In favourable conditions *Euglena* multiplies rapidly by a process of LONGITUDINAL BINARY FISSION. The nucleus undergoes mitosis when numerous chromosomes can be observed. During the formation of daughter nuclei other organelles are duplicated. The cytoplasm then undergoes longitudinal cleavage so that two identical daughter cells are formed. Encystment of the cell is an infrequent event during which the flagella are lost and the cell secretes a mucilagenous outer cover. Fission may occur within such cysts with the formation of numerous daughter cells which develop flagella when the cyst bursts.

4.4 CLASS *CILIATA*

The *Ciliata* is a large group of specialized protozoa. A large number of cilia, two or more kinds of nuclei per cell, a complex reproductive mechanism and other features unique for this level of development make them distinct from the other protozoan classes.

Paramecium is found in fresh water enriched by decaying vegetation, and can be easily cultured in the laboratory in a hay infusion (Chapter 11).

(a) Structure

Fig. 4.8   Structure of *Paramecium*

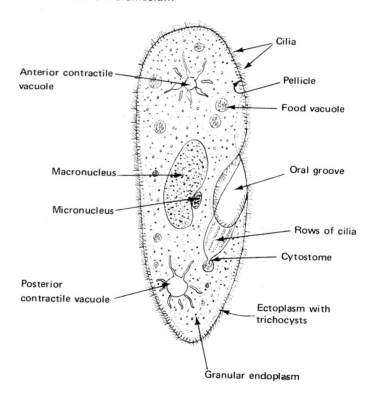

The cell body is elongate with a blunt anterior and a tapering posterior (Fig. 4.8). The protoplast is enclosed in a semi-rigid PELLICLE perforated by many CILIA arranged in longitudinal rows. Each cilium arises from a basal body or KINETOSOME, from the base of which fibrils pass to adjacent cilia. The complex network of fibrils constitutes the KINETODESMATA which is thought to maintain the shape and rigidity of the cell. Inside the pellicle the body is divided into a clear, outer ectoplasm and an inner granular endoplasm. The TRICHOCYSTS are specialized organelles situated in the ectoplasm. Their contents may be discharged outside the cell as fine threads to serve as a means of attachment during feeding or as weapons of defence. From the anterior end of the body a shallow ciliated ORAL GROOVE extends backwards. It narrows to form a short gullet at the end of which the mouth, or CYTOSTOME, opens into the endoplasm. To one side of the gullet is another pellicular modification, the CYTOPROCT (cell anus), where solid wastes are forcibly removed by vacuolar activity. In the endoplasm are FOOD VACUOLES, two CONTRACTILE VACUOLES with their associated canal systems for osmo-regulation, and two nuclei. One of the nuclei is a small rounded MICRONUCLEUS. It is partly concealed by a larger MACRONUCLEUS.

(b) Movement

The cilia, which have the $9+2$ microfibril organization typical of eukaryote cells, are the organelles of locomotion. They beat in a coordinated and regular manner called metachronal rhythm and the cell moves rapidly in a fairly straight path while rotating on its longitudinal axis. If *Paramecium* meets an obstruction or some unfavourable chemical stimulus, the beat of the cilia is reversed. The animal retreats a short distance, changes direction, then moves forward once again.

As with flagella, it is assumed that the necessary energy for movement comes from ATP and this substance has been found in the cilia. Precisely how the ciliary beat is controlled is not clear.

(c) Nutrition

A wide variety of micro-organisms, including other smaller protozoa, are swept into the oral groove and down to the gullet by ciliary action. At the cytostome, food vacuoles are pinched off and migrate through the endoplasm along a fixed path (Fig. 4.9). Using indicators such as Congo red, it is possible to show a change in pH in the vacuole contents from acid to alkaline as digestion occurs. Finally, the food vacuole meets the cytoproct vacuole where indigestible residues are discharged, the soluble products of digestion having already been absorbed by the cytoplasm from the vacuole.

Fig. 4.9   Path of food vacuoles in *Paramecium*

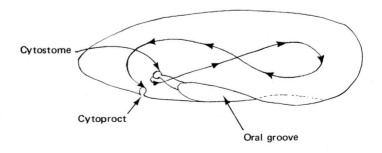

Cytostome

Cytoproct

Oral groove

(d) Reproduction

Paramecium reproduces both asexually and sexually. In the asexual process of BINARY FISSION, the micronucleus divides by mitosis into daughter nuclei which move to opposite ends of the cell. The macronucleus, having increased in size, divides by fission. When a second gullet and contractile vacuoles have been formed the cell divides transversely. By the time the two daughter cells separate they each contain a complete set of organelles. Under favourable conditions binary fission may occur several times per day, producing large numbers of identical progeny or CLONES.

Sexual reproduction is by a process of CONJUGATION which will only take place between compatible mating types of the same species of *Paramecium* (Fig. 4.10). Unlike the asexual

Fig. 4.10   Conjugation in *Paramecium caudatum*

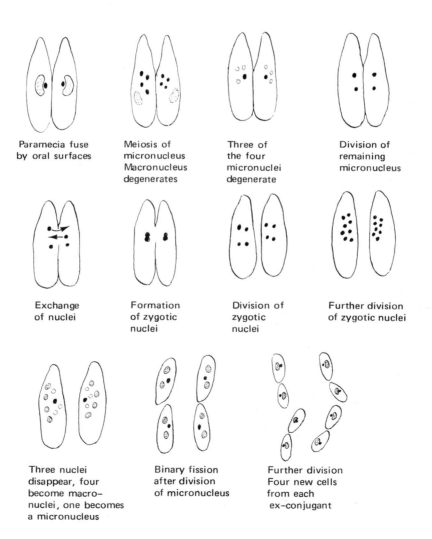

42

process which results in genetically identical offspring, conjugation produces progeny with a variety of genotypes as a result of meiosis and nuclear transfer. Self-fertilization or AUTOGAMY in which conjugation occurs within a single individual is a modification of the sexual process which sometimes occurs (Fig. 4.11). This process is obviously less important as a means of producing genetic change.

Fig. 4.11   Autogamy

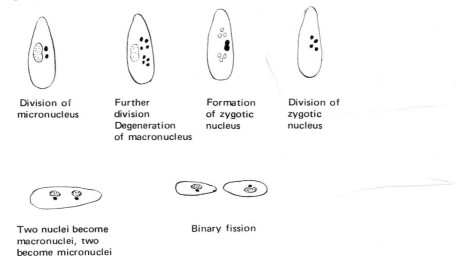

Division of
micronucleus

Further
division
Degeneration
of macronucleus

Formation
of zygotic
nucleus

Division of
zygotic
nucleus

Two nuclei become
macronuclei, two
become micronuclei

Binary fission

The major site of genetic material in *Paramecium* is undoubtedly the micronucleus but in addition some inherited characteristics are controlled by self-replicating particles found in the cytoplasm. An example of such PLASMAGENES is seen in the inheritance of kappa particles in *P. aurelia*. Cells containing kappa particles are so called killers due to the production of toxic substances to which they are resistant. Sensitive strains lack such particles and may be destroyed following the process of conjugation when cytoplasm containing the kappa material is transferred. It is now thought that kappa particles are endosymbiotic bacteria.

4.5 CLASS *SPOROZOA*

This major division of the phylum *Protozoa* contains a large and varied collection of species with little in common apart from a relatively simple cell organization and a complex life-cycle associated with a parasitic mode of life. Unlike other protozoan groups, the *Sporozoa* have no obvious organelles for locomotion.

Sporozoans occur as animal parasites in both invertebrate and vertebrate hosts but it is their occurrence as endoparasites of man and domesticated animals which is most important to us. Malaria in man, probably the most important human disease caused by a pathogenic micro-organism, and coccidiosis in poultry are two examples of disease caused by sporozoans.

(a) Structure of *Plasmodium* and its life-cycle

The causative organism of malaria belongs to the subclass *Telosporidea* in which a complex life-cycle involving SPOROGONY (multiple fission) of a zygote is followed by SCHIZOGONY of asexually produced cells. These stages of multiplication in the life-cycle occur in two distinct hosts, an *Anopheles* mosquito and a warm blooded vertebrate such as man (Fig. 4.12). The production of enormous numbers of offspring reflects the hazards of a life-cycle involving separate hosts.

Different species of *Plasmodium* can be recognized from the time interval between successive bouts of fever in the human host which mark the duration of the erythrocytic cycle. Regular 48-hour cycles of fever indicate benign tertian malaria caused by *P. vivax*. A 72-hour cycle is quartan (fourth day) malaria in which *P. malariae* is the pathogen. Malignant tertian malaria is caused by *P. falciparum* in which the asexual reproductive cycle is irregular.

(b) Control measures

Malaria has been a scourge of mankind for centuries. The disease reached its highest incidence in the 1940's with 350 million people suffering from malaria and 3 million deaths annually. Successful control of malaria can only be brought about by a combination of methods. Once the life-cycle of the parasite had been established steps could be taken to break it in its weakest place; namely when transfer from host to host occurs. Stop people from being bitten by the mosquito or remove the mosquito from the immediate vicinity of houses, and the life-cycle of *Plasmodium* is also stopped. The mosquito vector has to be eliminated and the protozoan killed inside the human host.

The vector has aquatic larvae and pupae which are suspended from the water surface by spiracles through which gaseous oxygen is breathed. Measures such as draining off, oiling or poisoning the water to kill the larvae and pupae are highly effective. Certain fish of the carp family which feed on the larvae and pupae can be introduced into the waters likely to be used by mosquitoes for breeding. Unfortunately the sheer size of such projects is against man. Mosquitoes can breed in inaccessible swamp grounds making it very difficult to organize eradication projects. Human activities in road building, irrigation schemes, clearing forests and the like produce temporary pools and puddles which make ideal breeding quarters for certain species of *Anopheles*.

The use of insecticides such as DDT to destroy the adult mosquito has proved effective. Spraying of dwellings leaves a toxic residue on walls and ceilings where mosquitoes rest. Problems of increased resistance to insecticides means that a continual search must be made for new and more powerful chemicals. Preventing healthy people from being bitten by mosquitoes when living in malarial regions is an almost impossible task despite the use of repellant creams and mosquito nets. However, a number of drugs are available to give some protection if the parasite is introduced into a human by the vector. Chemicals allied to quinine such as quinacrine and chloroquine can be taken orally to cure the disease. They stop schizogony in the red blood cells but have little effect on the persistent pre-erythrocytic stages of *P. vivax* in the liver. Quinine-based drugs also provide some immunity to infection

Fig. 4.12 Life cycle of *Plasmodium vivax*

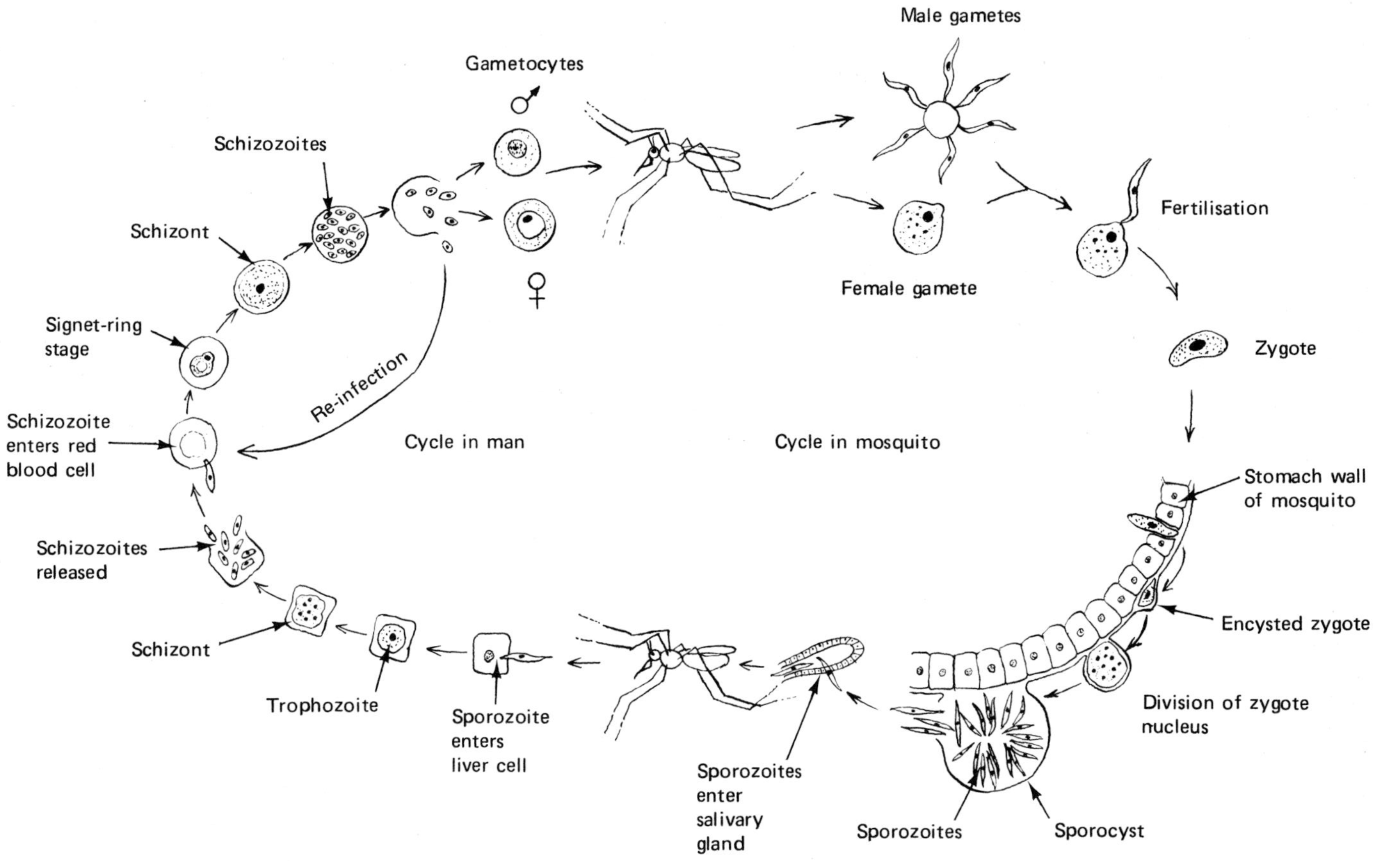

if they are taken regularly.

The World Health Organization has spent vast sums of money since 1955 on campaigns for malaria eradication. The early signs were that malaria could be conquered within 10 years. In India alone the number of cases of malaria dropped from 100 million in 1952 to just 60,000 in 1962. However by 1976 the disease had shown a dramatic comeback with over 6 million Indian victims. Similar reports have come from other countries and today malaria afflicts over 200 million people throughout the world.

The main reason for the resurgence of malaria is the appearance of insecticide-resistant mosquitoes. Another factor is the tolerance of several species of *Plasmodium* to standard anti-malaria drugs. Too little attention has possibly been paid to the ways in which man has altered the ecology of malarious countries and in so doing has provided extensive habitats such as paddy-fields in which mosquitoes breed. In China mass education and active participation of the population in eliminating breeding grounds of the mosquito have achieved considerable success in controlling malaria. Perhaps this course of action will now be followed by other countries where malaria is indigenous. There may be no other effective alternative.

5 VIRUSES

Virology is the youngest of the branches of microbiology. It has its origin in the last decade of the nineteenth century when Iwanowski, a Russian scientist, showed that tobacco mosaic disease was caused by an agent capable of passing through a filter that could retain bacteria. Within the next ten years filterable agents were found to be the cause of foot and mouth disease of cattle and yellow fever in humans. In 1915 viruses which grow on bacteria were discovered and these were later named bacteriophages ('phages). However, it was not until the 1930's that there was any information available on the composition of viruses when they were shown to consist of nucleoprotein. Attempts at seeing these infective agents up to then were thwarted by the limited resolution of the light microscope. It was only during the 1940's when the electron microscope was sufficiently refined and in general use that viruses could be seen for the very first time.

5.1 STRUCTURE AND SIZE

Most viruses consist of a protein coat or CAPSID which encloses the nucleic acid, either DNA or RNA. The capsid is made up of a number of sub-units or CAPSOMERES. Some viruses possess in addition an ENVELOPE composed of carbohydrate or lipo-protein while a few viruses contain one or two enzymes. The individual virus or VIRION is thus much less complicated in structure than the prokaryotic bacteria and blue–green algae or the eukaryotic fungi and protozoa. They are sometimes described as AKARYOTIC because of their extreme simplicity which reaches its limits in the potato tuber spindle virus where even a protein coat is absent.

On a morphological basis it is possible to distinguish three main viral shapes (Fig. 5.1, Plate 5.1).

(a) Helices

Tobacco mosaic virus (TMV) is a typical member of this group. In this species the conical capsomeres are arranged helically in the capsid in which the RNA is embedded. The virions appear as hollow rods at low magnification, a shape also observed in some 'phages.

In other viruses the helical capsid is coiled and contained within an envelope; for example, myxoviruses, the cause of influenza, measles and mumps.

(b) Polyhedrons

In polyhedral viruses the capsomeres are arranged into a many-sided shape, the most common an icosahedron having twenty faces, each face being an equilateral triangle. The number of capsomeres making up each side varies from virus to virus and there is a considerable range of size. This group includes enveloped forms; for example, poliomyelitis virus, and Herpes virus, the cause of cold sores, as well as naked forms such as adenoviruses which infect adenoids and tonsils, polyoma viruses

47

which include the human wart virus, and some bacteriophages. The nucleic acid is found inside the capsid, possibly sandwiched between the latter and an inner protein coat. Some members of this group contain DNA, others possess RNA.

Fig. 5.1   Viral shapes

HELICES

Tobacco mosaic virus   Myxovirus

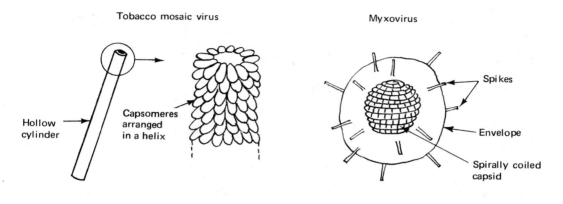

POLYHEDRONS (capsomeres shown on one face only)

Adenovirus   Herpes virus

COMPLEX VIRUSES

Bacteriophage   Vaccinia

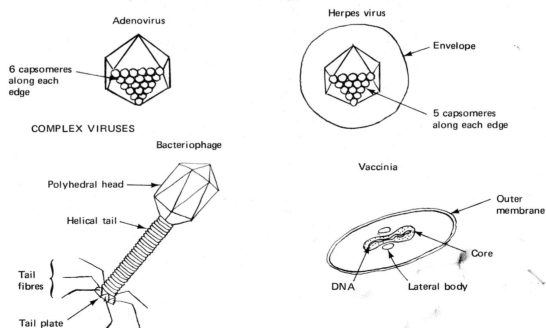

Plate 5.1 Electronmicrographs of viruses:

1. Bacteriophage,
x 200,000

2. Influenza virus,
x 230,000

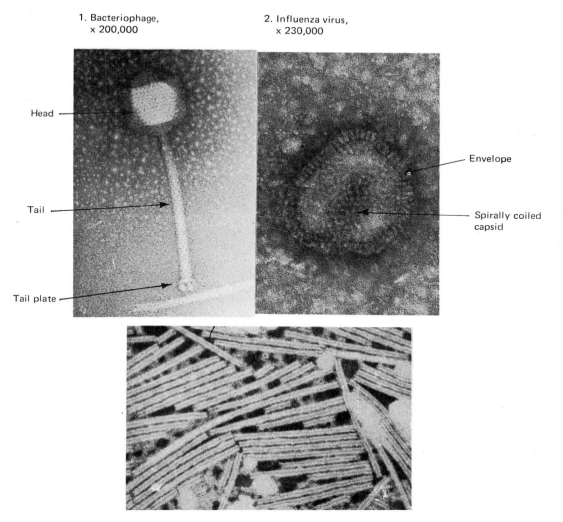

Head

Tail

Tail plate

Envelope

Spirally coiled
capsid

3. Tobacco mosaic virus, x 100,000

(c) Complex viruses

The best known of the complex viruses are the bacteriophages which have an icosahedral head, and a helical tail. In addition they possess a tail collar and a tail-plate from which long filaments project. DNA is present as a single helix inside the head. 'Phages of basically the same shape have also been found in blue—green algae. They are called cyanophages.

Pox viruses, vaccinia, also display a complex symmetry but which is quite different from that of 'phages. These have an outer double membrane, an inner elongated biconcave core containing DNA and two lateral bodies whose function is unknown.

The largest virions such as those of polyhedral psittacosis virus measure 400—500 mμ in diameter and are just visible under the light microscope. The limit of resolution of this instrument is 200 mμ. In practice the electron microscope has a resolution limit of 1 mμ and with this instrument polyhedral viruses as small as 8–12 mμ in diameter, for example foot and mouth virus, have been observed. It is conceivable that viruses even smaller than this are in existence.

Table 5.1 Dimensions of virions

SYMMETRY	NAME	DIMENSIONS
Coiled helix	Myxovirus	80–120 mμ diameter
Simple helix	Tobacco mosaic virus	300 mμ long × 15 mμ wide
Polyhedron	Adenovirus	70–90 mμ diameter
Polyhedron	Herpes virus	120–180 mμ diameter
Polyhedron	Poliomyelitis virus	28 mμ diameter
Polyhedron	Psittacosis virus	450 mμ diameter
Complex	Cyanophage	Head: 60 mμ long × 60 mμ wide Tail: 20 mμ long × 15 mμ wide
Complex	T2 coliphage	Head: 95 mμ long × 65 mμ wide Tail: 100 mμ long × 25 mμ wide
Complex	Vaccinia virus	300 mμ long

5.2 CULTIVATION AND PURIFICATION OF VIRUSES

Viruses are obligate parasites and must be cultured in living cells. The only method available for growing animal viruses at one time was to inoculate them into susceptible animals such as mice. This method is still used for difficult types. It was later discovered that some viruses could be cultured in fertile hen's eggs where the embryo, embryonic membranes and yolk sac provide ideal conditions for viral growth.

A small hole is cut into the egg shell using a dentist's drill and a suspension of virions is then injected into a suitable part of the egg (Fig. 5.2). Vaccinia virus, the cause of smallpox, grows well in the chorio-allantoic membrane while myxovirus, the cause of influenza and mumps, multiplies in the amniotic and allantoic cavities. Vaccines which are used to give immunity against influenza can be produced from the amniotic fluid. The liquid is extracted aseptically and centrifuged at high speed. Eventually a pure culture of virions is obtained in suspension. This is inactivated by treatment with formaldehyde and finally suspended in neutral buffered saline containing a bactericide. Because of their host specificity, not all animal viruses can be cultured in hen's eggs. Recently the practice of tissue culture has been widely used to cultivate some viruses such as the poliomyelitis virus, which is grown in cultured kidney cells of the Rhesus monkey. The kidneys are aseptically removed, minced and then suspended in a solution of the enzyme trypsin which releases individual cells into the medium. Small quantities of the cells are transferred to a growth medium in large bottles where they multiply to form a layer one cell thick which adheres to the surface of the glass. This is then inoculated with a suspension of virions which grow in some of the cells to form opaque areas known as plaques. Infected cells release virions into the medium which can then be purified as described for flu virus. It is in this way that the inactivated Salk-type poliomyelitis vaccine is produced. Attenuated Sabin-type vaccine is made by using mild strains of the virus and there is no inactivation stage.

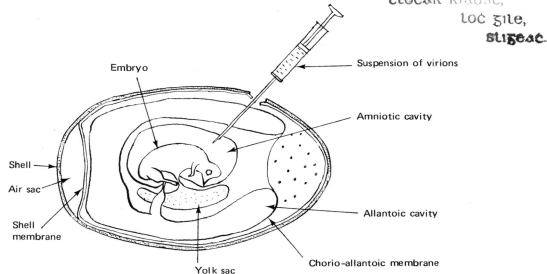

Fig. 5.2   L.S. fertile hen's egg at 10-14 days showing injection with myxovirus

Embryo

Suspension of virions

Amniotic cavity

Shell

Air sac

Shell membrane

Allantoic cavity

Chorio-allantoic membrane

Yolk sac

Plant viruses can be cultured by inoculating susceptible host plants. This is done by rubbing a leaf or young stem with an abrasive powder or scratching the outer tissue and then applying either a pure culture of the virus or the sap removed from an infected plant. Such methods are known as sap-inoculation. The practice of tissue culture has also been used for growing many plant viruses.

Infected plant tissue is minced and the sap expressed under force. The sap is then centrifuged at low speed to remove cell debris. A number of techniques are available to purify the virus:

(*i*) Alternate high and low speed centrifugation.
(*ii*) Precipitation using ammonium sulphate or 90% ethanol followed by differential centrifugation.
(*iii*) Gel-filtration in which the suspension is poured down a 1% agar column which retains small non-infective particles but allows infective virions to pass through.

5.3 VIRUS REPLICATION

(a) The lytic cycle

The bacteriophages have been the subject of many studies of viral reproduction. T2 coliphage becomes attached to a specific site on the wall of an *Escherichia coli* cell, rather like the attraction between enzyme and substrate molecules. The wall of the bacterium is then softened by the secretion of a lysozyme enzyme from the tail plate (Fig. 5.3). Now the tail contracts (Plate 5.2) and the DNA is squeezed out of the head into the host cell. It has been possible to separate the capsid and nucleic acids of many viruses and it has been found that the nucleic acid alone is capable of infecting a host.

There now follows a short LATENT or ECLIPSE phase during which little can be seen to happen. During this period the viral DNA represses the metabolic reactions of the host and directs the synthetic processes towards the production of new 'phage particles. Initially the messenger RNA of the bacterium is used to produce 'phage-coded enzymes essential for the replication of 'phage DNA later on. This is followed by the synthesis of 'phage capsomeres (protein) and 'phage DNA. These are then assembled into new virions, the head being put together first and finally the tail. Ultimately a lysozyme-like enzyme is made which hydrolyses the mucopeptide of the bacterial wall. This results in the osmotic uptake of water causing the bacterium to undergo lysis.

Fig. 5.3   Entry of coliphage and the lytic cycle

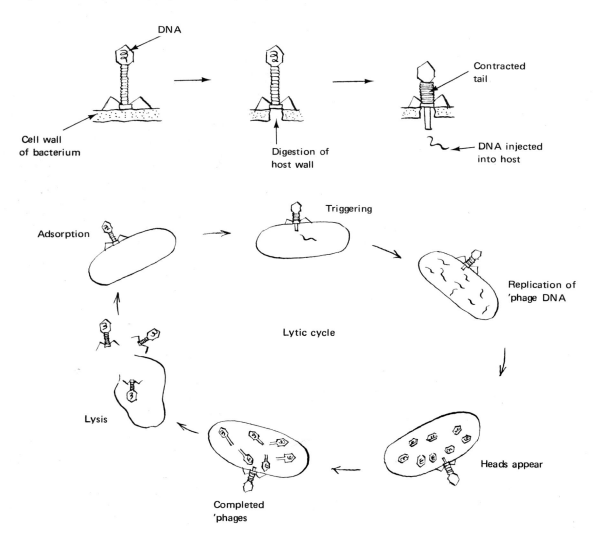

Plate 5.2 Triggered bacteriophage, x 200,000

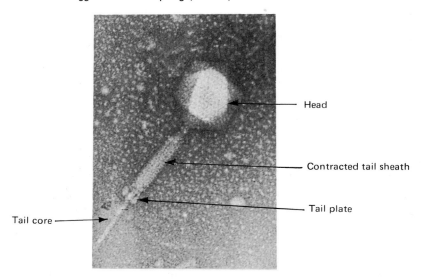

Head

Contracted tail sheath

Tail plate

Tail core

(b) Growth curve for virions

Because of their minute size it is obviously impossible to visually count the number of virions released from a host cell. Instead a simple cultural method is used. Bacteria, or other suitable host cells, are incubated with a 'phage inoculum and then samples are plated out in a nutrient agar medium at staggered intervals of time. The bacteria grow to form a confluent plate or lawn. Any bacterium infected with a 'phage fails to grow leaving a hole or plaque in the lawn. By counting the number of plaques appearing on each plate a graph can be constructed which is shown in Fig. 5.4. At first the number

Fig. 5.4   One-step growth curve of T2 coliphage

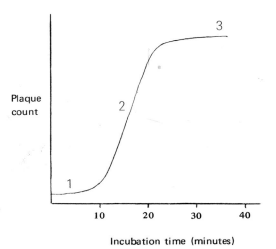

Plaque count

Incubation time (minutes)

of plaques remains small but constant. This is the LATENT or ECLIPSE PHASE (1) and lasts for about ten minutes. There is then a rapid increase in plaque numbers over the next ten minutes called the RISE PERIOD (2), after which the number again settles down to a consistently high value (3). The number of plaques at this stage is known as the BURST SIZE and represents the factor by which the viral number has risen during the experimental time. This is the average number of 'phage particles released from a single infected cell. Such a pattern of growth is known as the ONE-STEP growth curve.

5.4 LYSOGENY

The lytic cycle described earlier is displayed by all virulent 'phages but there are instances when the virus becomes avirulent or TEMPERATE. Instead of causing the death of its host cell 'phage DNA occasionally becomes attached to the bacterial chromosome and is known as PROVIRUS. It replicates synchronously with *E. coli* DNA prior to fission and is then passed on to the daughter cells (Fig. 5.5). The presence of provirus can cause a bacterium to develop phenotypic features it had not previously possessed; for example, the diphtheria bacillus *Corynebacterium diphtheriae* is only able to produce toxin if it contains specific types of provirus. The disease symptoms in a susceptible human host are entirely due to the toxin being released from the bacillus. In addition, bacteria containing provirus appear to develop immunity against attack by the same strain of virus.

Fig. 5.5   The lysogenic cycle

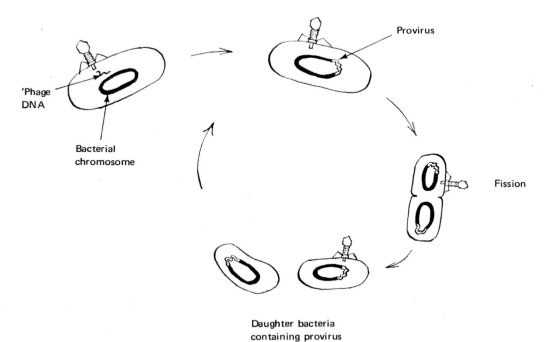

'Phage DNA

Bacterial chromosome

Provirus

Fission

Daughter bacteria containing provirus

It is thought that the provirus possesses a gene which inhibits its release into the cytoplasm. From time to time the 'phage may be freed from the bacterial chromosome and on entering the cytoplasm it reverts to a virulent form. It is not clear what causes this but the condition can be induced with agents such as ultra-violet light so that it is likely to be a mutagenic change. On lysis the 'phage is now able to infect non-lysogenized bacteria upon which it may confer a new physiological or structural feature. This phenomenon is termed TRANSDUCTION and because many bacteria appear to contain provirus it is probably an important means of providing variation among bacterial populations which can be tested in the environment.

6 FUNGI

At least one hundred thousand species of fungi are known to man. Many, for example yeasts and moulds, can only be seen under the microscope. Others such as mushrooms, toadstools and puffballs are organisms of substantial size. Most species are useful in one way or another, and play an important role in ecosystems because of their ability to help in the recycling of nutrients and energy within the biosphere. Soil fertility for example depends on the breaking-down activities of many species of terrestrial fungi. Together with bacteria, actinomycetes and the soil fauna, they reduce dead organic matter to simple soluble minerals and gases which can be used again by higher plants. Unfortunately the same organisms can sometimes cause serious economic losses when they colonize and spoil products such as food, wood, paper, leather, textiles and paint.

Plate 6.1 Types of fungi

| 1. Sporangium of *Mucor*, x 200. | 2. Zygospores of *Mucor*, x 200. | 3. Sporangium of *Saprolegnia*, x 200. | 4. Oogonium of *Saprolegnia*, x 400. |

Phycomycetes

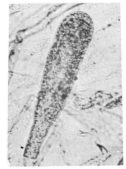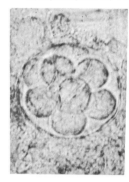

| 5. *Schizosaccharomyces* fission and asci, × 1,000 | 6. Conidia of *Penicillium*, x 200. | 7. Conidia of *Aspergillus*, x 200. | 8. Asci of *Sordaria*, x 200. |

Ascomycetes

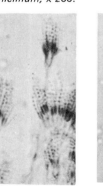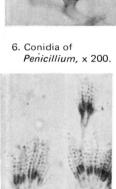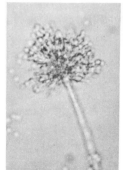

56

Plate 6.1 (continuation)

Basidiomycetes

9. Toadstool
Collybia, x $\frac{1}{2}$

10. Basidia of
Coprinus, x 400.

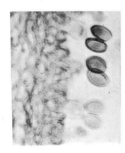

11. Bracket-fungus
Polyporus, x$\frac{1}{25}$

12. Teleutospores
of *Puccinia*, x 100.

Fungi Imperfecti

13. Conidia
of *Fusarium*, x 200.

14. Conidia
of *Monilia*, x 100.

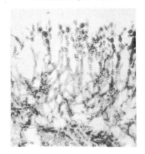

15. Pycnidium
of *Phoma*, x 200.

Other species cause death or disease of living organisms. In spite of the widespread use of fungicides pathogenic fungi are still responsible for considerable loss of yield in many crop plants. Man and domesticated animals are also infected by fungi; the more common mammalian pathogens cause skin infections such as ringworm and athlete's foot.

Some of the biochemical activities of fungi have long been exploited by man. For centuries yeasts have been used in bread-, beer- and wine-making, and moulds to improve the flavour of certain cheeses. Fleming's discovery in the 1920's of the production of penicillin by a mould stimulated interest in the physiology of the fungi and in the past decade one of the more remarkable discoveries has been the transformation of steroids by fungi into drugs used in birth control and in the treatment of arthritis.

Although their nutritional value is probably small some of the larger fungi such as mushrooms, truffles and puffballs have been prized as gastronomic delicacies since Greek and Roman times. At the same time there are regular reports of deaths where poisonous species have been eaten by mistake. Unicellular fungi such as yeasts are much more nourishing and their efficiency at producing protein is quite phenomenal, fifty times as rapid as that of a bullock for example. As the world demand for protein grows fungal protein could well be used to feed future generations, and experimental industrial plants have already been set up with this in mind (Chapter 8).

6.1 CLASSIFICATION OF FUNGI

Differences in anatomy and reproduction form the basis of fungal classification (Table 6.1):

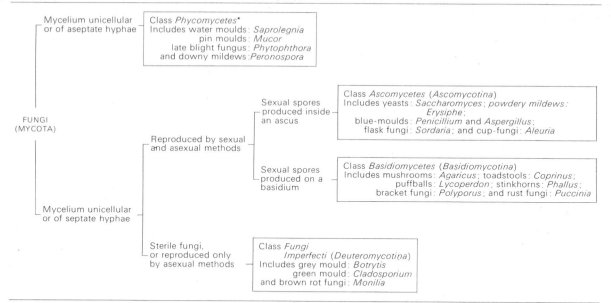

* Recent classifications divide the *Phycomycetes* into the *Mastigomycotina* (zoosporic fungi) and the *Zygomycotina* (zygosporic fungi) which are equal in rank to the *Ascomycotina, Basidiomycotina* and *Deuteromycotina.*

6.2 STRUCTURE

Fungi usually consist of fine, branched threads or HYPHAE which form an entangled, cottony mass known as a MYCELIUM. A hypha may measure from 0.5 μm to 1.0 mm in diameter, according to species, and will vary considerably in length. In the *Phycomycetes* there are no cross walls (septa) inside the hyphae except where reproductive organs are cut off or where damage occurs. (Fig. 6.1). The mycelium is said to be ASEPTATE or COENOCYTIC and there is a single multinucleate protoplast enclosed within a common cell wall. SEPTATE hyphae are found in the *Ascomycetes* and *Basidiomycetes* where the protoplast between two successive septa may have one or several nuclei. In some basidiomycete fungi lateral bulges (clamp connections) develop on the hyphae at certain stages in the life history.

Even the smallest fungus such as a yeast (*Saccharomyces*) which is unicellular can be readily distinguished from a bacterium because fungal ultrastructure is similar to that of higher plants (Fig. 6.2). There is a distinct NUCLEUS surrounded by a porous nuclear membrane and chromosomes appear during nuclear division. MITOCHONDRIA, with their characteristic invaginated inner membranes (cristae), are scattered throughout the cytoplasm in which is found an extensive ENDOPLASMIC RETICULUM. These organelles have not been observed in bacterial cells. RIBOSOMES float freely in the cytoplasm or are attached to membranes of the endoplasmic reticulum. Vacuoles containing storage materials such as glycogen, a polysaccharide

Fig. 6.1   Aseptate and septate hyphae

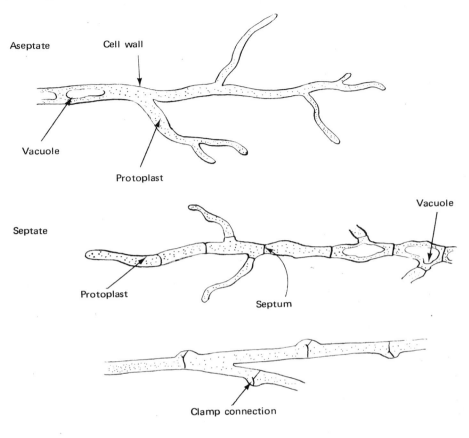

Plate 6.2 Ultrastructure of fungal cell wall, x 20,000.
The hypha has been partly digested with chitinase enzymes.

similar in structure to starch, lipid globules and volutin, a metaphosphate polymer, are common. The protoplast is bounded by a semi-permeable PLASMA MEMBRANE, which is a phospholipid unit membrane. It is enclosed in a rigid, permeable wall, which, like that of higher plants, consists of a fibrillar network supported by an amorphous matrix (Plate 6.2). The yeast cell wall is composed largely of polymers of mannose (mannans) and glucose (glucans) but in most fungal walls CHITIN (Fig. 6.3), a polymer of N-acetylglucosamine, is the major fibrous constituent.

Fig. 6.2   Electronmicrograph drawing of a thin section of a cell of *Saccharomyces*

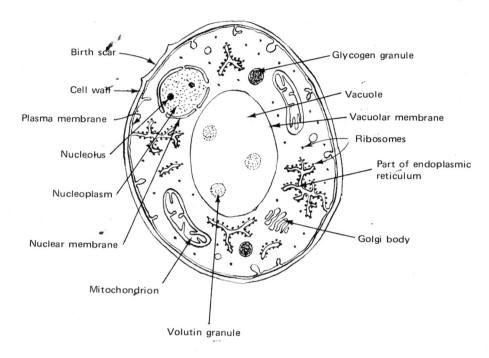

Fig. 6.3   Chitin, a polymer of 1:4 N'acetylglucosamine

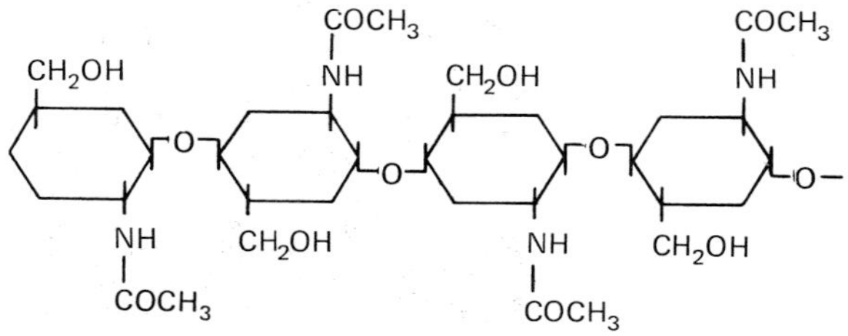

60

The ultrastructure of filamentous fungi is basically similar to that of yeasts (Plate 6.3). At the tips of young hyphae the dense protoplasm is packed with ribosomes, mitochondria and small vesicles. It is here that new wall material, organelles and enzymes are rapidly synthesized. Older parts of hyphae which are usually metabolically inactive often have conspicuous vacuoles in the cytoplasm.

Plate 6.3 Thin section of hypha of *Aspergillus nidulans,* x 10,000.

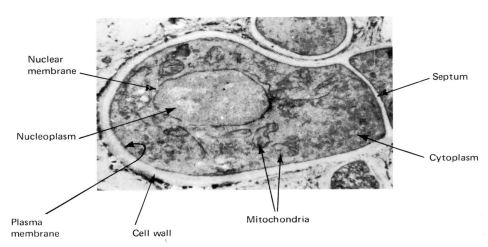

In septate species pores are present in the cross walls which allow substances in solution and suspended solids including nuclei to move freely from cell to cell (Plate 6.4).

Plate 6.4 Septum of *Aspergillus nidulans,* x 15,000.

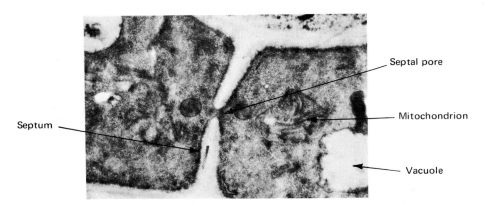

6.3 VEGETATIVE MORPHOLOGY

The shape of the fungal body will to some extent depend on the conditions under which growth occurs. When a fungal spore is inoculated on to a solid nutrient medium it swells and produces a germ tube. Lateral branches grow from this tube and soon a network of hyphae appears. Each branch grows at its tip, divides repeatedly (Fig. 6.4) and continually explores areas of fresh nutrient while growing away from its own waste products (negative chemotropism). The hyphae hardly penetrate the medium probably due to lack of enough oxygen in the deeper layers. In most cases a fine turf-like or fluffy colony circular in outline is formed within a few days (Plate 6.5).

Fig. 6.4   Stages in growth of a fungal colony.

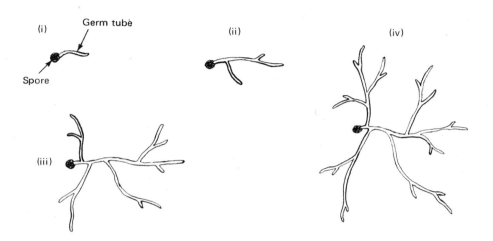

Fungi grown in a stationary liquid medium form a floating felt or mat. When the medium is oxygenated by pumping in sterile air or by continually shaking the culture flasks, a number of small, more or less spherical masses of mycelium known as pellets appear (Plate 6.5).

Plate 6.5   Vegetative morphology of fungi:

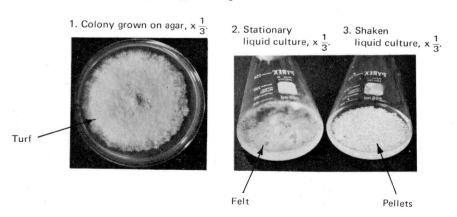

1. Colony grown on agar, x $\frac{1}{3}$.

2. Stationary liquid culture, x $\frac{1}{3}$.

3. Shaken liquid culture, x $\frac{1}{3}$.

Turf

Felt

Pellets

6.4 REPRODUCTION AND SURVIVAL

After colonizing a fresh substrate, most fungi enter a phase of vegetative growth. There then follows a phase of reproduction, asexual at first and later but not always sexual. Some species reproduce only by sexual means. Non-sexual methods are probably more important than sexual for spreading the species. Asexual spores can be produced in large quantities several times during a season. Because they do not involve nuclear fusion, asexual methods almost invariably give rise to progeny of similar genotype to that of their parents.

(a) Asexual reproduction

Fig. 6.5   Production and release of asexual spores by phycomycete fungi

(i) *Saprolegnia*

(ii) *Mucor*

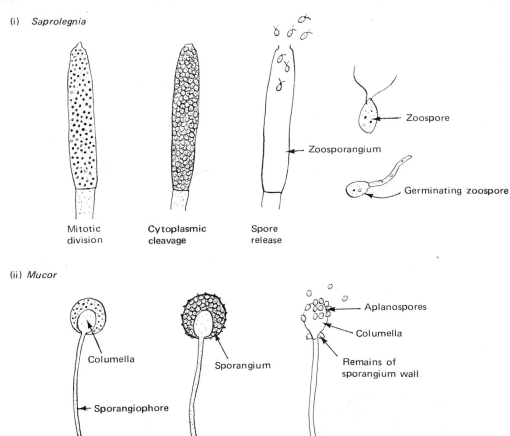

The characteristic asexual propagule is the SPORE, but the way in which spores are produced differs from species to species. *Phycomycetes*. Water moulds such as *Saprolegnia* produce motile unicellular spores called ZOOSPORES inside a spherical or tubular sac termed a ZOOSPORANGIUM. The flagellate spores arise following mitotic nuclear division and cytoplasmic cleavage within the sac. On release into the water the spores are

chemotactically attracted to fresh substrates. Mucoraceous
moulds form non-motile unicellular spores known as APLANO-
SPORES inside spherical SPORANGIA (sing. SPORANGIUM) at the tips
of simple or branched aerial hyphae called SPORANGIOPHORES. In
some species the sporangium wall is deliquescent and dissolves
when ripe. The spores are dispersed by means of air currents
(Fig. 6.5). Some phycomycete fungi such as *Phytophthora
infestans* display adaptations to both terrestrial and aquatic
habitats in that the way in which the asexual propagules
germinate is determined by environmental conditions (Chapter
8).

Ascomycetes. The most common asexual reproductive unit in
this class is the CONIDIUM (pl. CONIDIA). Conidia are well adapted
for dispersal in terrestrial habitats. Some of the simplest conidia-
bearing structures are seen in the powdery mildews which are
pathogenic to higher plants. In *Erysiphe graminis* the conidiophore is a flask-shaped cell borne directly on the mycelium.
This elongates and cuts off chains of dry conidia which spread
by air currents. In *Aspergillus* and *Penicillium* the spores are
produced inside phialides and the conidiophores are more
complex and highly characteristic of the genera (Fig. 6.6).

Fig. 6.6   Production of conidia by ascomycete fungi

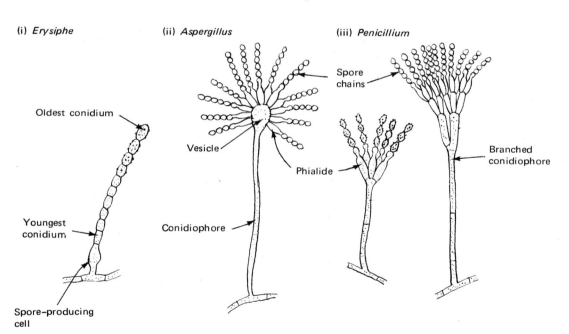

(i) *Erysiphe*   (ii) *Aspergillus*   (iii) *Penicillium*

Yeasts such as *Saccharomyces* multiply by BUDDING (Fig.
6.7). The nucleus does not undergo the usual mitotic
mechanism observed in other fungi. Instead the nuclear
membrane remains intact and chromosomes and a spindle
appear inside it. The membrane then undergoes constriction
and the nucleus is eventually pinched into two equal parts.
One of the nuclei, together with other organelles such as

mitochondria, then enters a bulge which develops at a point adjacent to the site of nuclear division. Cytoplasmic connection between the bud and parent cell is then broken followed by the laying down of new wall material.

Fig. 6.7   Budding in *Saccharomyces*

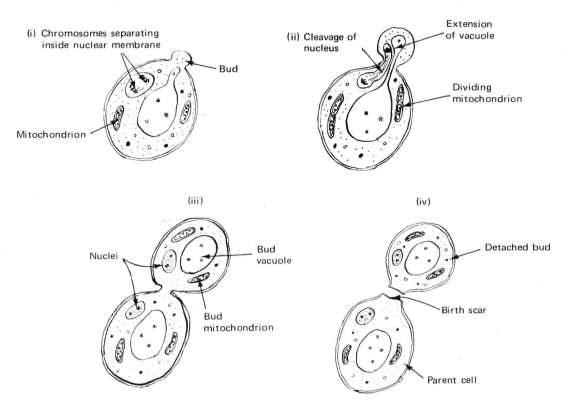

Basidiomycetes. Asexual sporulation is not a typical feature of basidiomycete fungi but a few species produce conidia-like spores. *Puccinia graminis,* the cause of black rust of wheat, has a complex life history during which several types of spores are formed. From early summer onwards tufts of uredospores appear on the leaves of infected plants (Plate 6.6). These spores are comparable to conidia and serve to spread the pathogen.

Fungi Imperfecti. The majority of species in this class produce CONIDIA and there is a great variety of spore-bearing structures. In some species such as *Phoma,* the conidia are borne inside flask-shaped organs (pycnidia) which are sunk below the surface of the substrate, but in most instances they arise from aerial conidiophores or even from unmodified hyphae. Those species of *Aspergillus* and *Penicillium* which appear to have lost the ability to reproduce sexually are often placed in this class together with many other dry-spored species; for example, *Botrytis cinerea* (grey mould of higher plants) and *Cladosporium*

65

herbarum (the spores of which are isolated from the atmosphere more frequently than those of any other fungus). Many soil-living imperfect fungi such as *Fusarium,* with its characteristic sickle-shaped conidia, produce slimy spores (Fig. 6.8).

Plate 6.6 T.S. leaf infected with *Puccinia,* x 200.

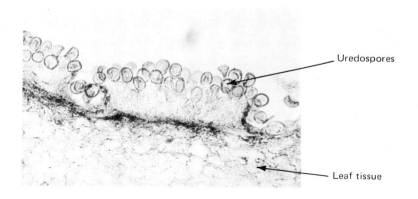

Uredospores

Leaf tissue

Fig. 6.8   Asexual sporulation in imperfect fungi

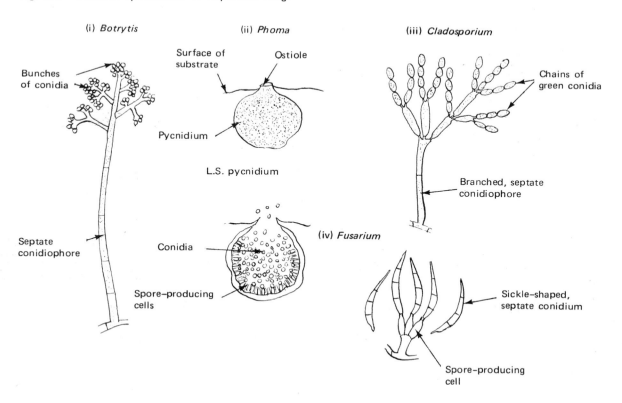

(i) *Botrytis*

Bunches of conidia

Septate conidiophore

(ii) *Phoma*

Surface of substrate

Ostiole

Pycnidium

L.S. pycnidium

Conidia

Spore–producing cells

(iii) *Cladosporium*

Chains of green conidia

Branched, septate conidiophore

(iv) *Fusarium*

Sickle–shaped, septate conidium

Spore-producing cell

(b) Sexual reproduction

Sexual reproduction involves the fusion of nuclei, often, but not always, from different strains of the same species, followed by meiosis at some stage in the life-cycle. Many fungi have evolved methods of ensuring that they outbreed and this gives scope for genetic variation which is hardly possible by the asexual process. At first glance it may be thought that the *Fungi Imperfecti* are an evolutionary backwater because of their lack of sexuality. However, new genotypes can arise in this group by mutation and from an unusual mitotic nuclear division in which crossing over is thought to occur (PARASEXUALITY).

Phycomycetes. The majority of water moulds such as *Saprolegnia* are HOMOTHALLIC (monoecious), that is, a mycelium derived from a single zoospore will give rise to both male and female unicellular sexual organs (GAMETANGIA) and self-fertilization takes place. The female organ, called an OOGONIUM,

Fig. 6.9   Sexual reproduction in phycomycete fungi

(i) *Saprolegnia*

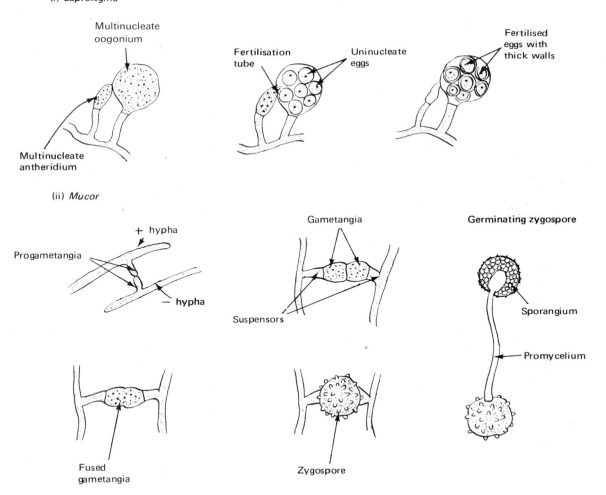

(ii) *Mucor*

at first contains a multi-nucleate protoplast which later becomes organized into five to ten uninucleate eggs. These are fertilized by male nuclei which enter the oogonium through a fertilization tube from an adjacent ANTHERIDIUM (Fig. 6.9). Each fertilized egg secretes a thick wall to become an OOSPORE which rests until conditions are favourable when it germinates to form a promycelium bearing a sporangium.

Most mucoraceous moulds are HETEROTHALLIC (dioecious) that is, a mycelium from a single sporangiospore fails to reproduce sexually unless it mates with another mycelium. Because the sexual organs on the two mycelia are alike in appearance (Fig. 6.9) it is impossible to designate one mycelium as male and the other as female. They are usually described as + and − mycelia. Within a few days of fertilization the + and − nuclei in the ZYGOSPORE fuse in pairs and meiosis occurs immediately. Following a period of rest a zygospore germinates to form a PROMYCELIUM bearing a sporangium which contains either + or − spores, not both. It is possible that only one diploid nucleus survives in the zygospore and when this undergoes meiosis it produces four nuclei (two with the + gene and two with the − gene). One of these two classes of nuclei degenerates, leaving the other to produce the nuclei of the spores by mitosis.

Ascomycetes. The characteristic product of sexual reproduction in the *Ascomycetes* is a sac-like ASCUS containing ASCOSPORES. In the heterothallic *Saccharomyces cerevisiae* (baker's yeast) conjugation between a pair of small cells containing haploid nuclei produces a larger diploid cell. Following meiotic division of the diploid nucleus four ascospores are formed. On release the spores grow into haploid cells (Fig. 6.10). Thus the haploid cells may be regarded as gametangia while the diploid cell becomes an ascus after its nucleus has divided meiotically.

Fig. 6.10   Life-cycle of *Saccharomyces cerevisiae*

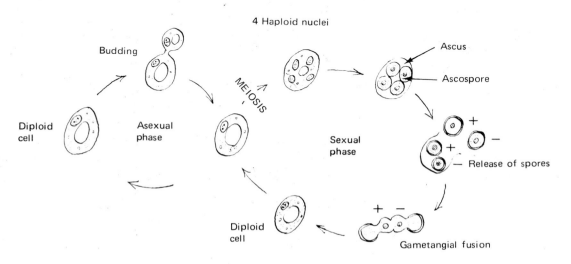

Fig. 6.11 Sexual reproduction in *Sordaria fimicola*

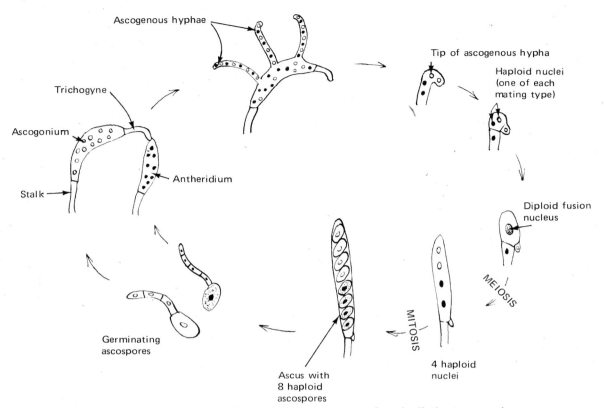

Ascogenous hyphae

Tip of ascogenous hypha

Haploid nuclei
(one of each
mating type)

Trichogyne

Ascogonium

Antheridium

Stalk

Diploid fusion
nucleus

MEIOSIS

MITOSIS

Germinating
ascospores

Ascus with
8 haploid
ascospores

4 haploid
nuclei

In most ascomycete fungi distinct sexual organs appear before conjugation (Fig. 6.11) and a protective structure, the ASCOCARP, grows from sterile hyphae which carry the developing asci. The ascocarp may completely enclose the asci so that the ascospores are released when it decays, as in *Penicillium wortmanni,* or it may possess a pore through which the spores are shot prior to dispersal as in *Sordaria fimicola.* A further modification, seen in the orange peel fungus, *Aleuria aurantia,* is a disc-shaped ascocarp bearing exposed asci from which the ascospores are forcibly ejected by a puffing action (Plate 6.7).

Plate 6.7 *Aleuria* ascocarp

1. Ascocarp growing
on a dead twig, x 1.

2. Thin section of
ascocarp, x 100.

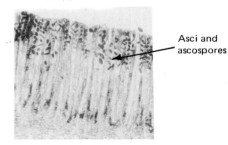

Asci and
ascospores

69

Basidiomycetes. Most basidiomycete fungi have no sexual organs at all and undifferentiated somatic cells often take over the function of gametangia (Fig. 6.12). As in the larger ascomycete fungi, nuclear fusion does not occur immediately after conjugation. From the product of hyphal fusion a secondary mycelium of binucleate cells with clamp connections arises, which gives rise to the characteristic toadstool fruiting body (BASIDIOCARP). Radiating gills (LAMELLAE) are found underneath the cap (PILEUS) and on the surface of the gills the spore-producing cells or BASIDIA, typical of this group of fungi, are formed. The development of a basidium is comparable to that of an ascus except that the basidiospores are EXOGENOUS (borne externally) unlike the ENDOGENOUS ascospores which remain within the ascus until dispersed. Some agarics have slight differences in basidiospore development from that outlined here. In the cultivated mushroom, *Agaricus bisporus* for example, each basidium forms two binucleate spores.

Fig. 6.12   Life-cycle of *Coprinus lagopus*

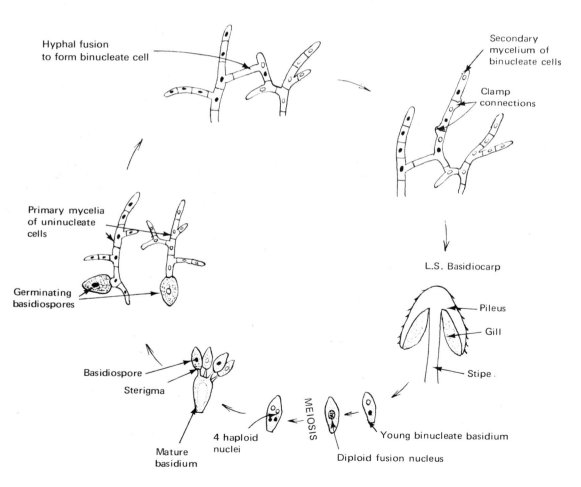

70

Various theories have been suggested to explain the mechanism of basidiospore release. Most mycologists agree that the spores are ejected forcibly over a short distance in a horizontal plane and then fall vertically between the gills. Geotropic responses of the stipe (negative) and gills (positive) maintain the gills in a vertical position so that the spores can fall freely from the pileus to be dispersed by air currents.

The forcible method of spore release is also used by bracket-fungi where the basidia are usually found lining vertical tubes on the lower surface of the basidicarp. Puff-balls rely largely on the pattering action of rain-drops to release their powdery spores from the ruptured, papery fruiting body while the sticky spores of stinkhorns are dispersed by insects which are attracted by the distinctive and unpleasant smell of the basidiocarp.

(c) Survival

When the energy and nutrients of a substrate have been exhausted or when other factors prevent continued growth fungi are faced with two alternatives if they are to survive. Either they must quickly find a new environment in which they can resume activity, or they can form resistant bodies which will remain alive until fresh substrates and/or improved conditions once again appear in the same place. Many of the asexual spores produced by fungi are probably best regarded

Fig. 6.13   Sclerotia

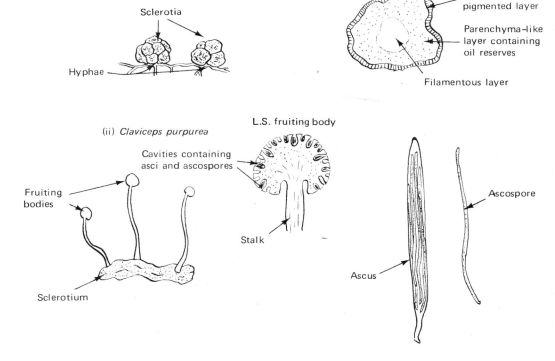

(i) *Botrytis cinerea*

L.S. sclerotium

Sclerotia

Hyphae

Thick-walled, pigmented layer

Parenchyma-like layer containing oil reserves

Filamentous layer

(ii) *Claviceps purpurea*

L.S. fruiting body

Cavities containing asci and ascospores

Fruiting bodies

Stalk

Sclerotium

Ascus

Ascospore

as bodies of dispersal because they do not usually survive for more than a few months. However, some species can produce propagules which remain viable for several years in unfavourable circumstances, although such bodies are often formed even when food is plentiful.

Botrytis cinerea (grey mould) forms small masses of tightly packed hyphae known as SCLEROTIA in which the mycelium in section looks like parenchyma tissue. The peripheral cells are thick-walled, closely packed and often pigmented (Fig. 6.13). They protect the underlying tissue, which contains reserve food, from desiccation and extremes of temperature. Sclerotia are also formed by certain plant parasites such as *Claviceps purpurea* (the cause of ergot in grasses and cereals). Here the survival bodies known as ergot, produced in diseased grain, fall to the ground and over-winter. The following spring they give rise to stalked fruiting structures containing ascospores (Fig. 6.13). This species is of medicinal interest as the sclerotia contain toxic alkaloids, which when eaten by humans cause convulsions, hallucinations and even death. Cattle and sheep abort their foetuses after eating ergot.

Other important asexual survival bodies are CHLAMYDOSPORES which are often formed inside a hypha. The contents of the cells enlarge, round off and secrete a thick wall which protects the dense oily cytoplasm of the spore. They appear to be more resistant to digestion by soil bacteria than are hyphae and conidia. Chlamydospores may also arise from cells of multi-cellular spores under the right conditions (Fig. 6.14).

Fig. 6.14 Chlamydospores of *Fusarium oxysporum*

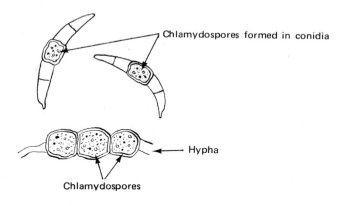

Chlamydospores formed in conidia

Hypha

Chlamydospores

Many fungi grow away from a substrate with few nutrients left. In *Armillaria mellea* (honey agaric), a parasite of forest trees, flattened rope-like strands or RHIZOMORPHS arise from the aggregation of large numbers of parallel hyphae (Fig. 6.15). The rhizomorph spreads the infection by growing from a food base such as a rotten tree stump to the roots or the trunks of healthy trees (Plate 6.8). Growth of rhizomorphs is both rapid (5–6 times faster than that of normal hyphae) and extensive (up to 10 m in length has been recorded).

72

Fig. 6.15   T.S. rhizomorph of *Armillaria mellea*

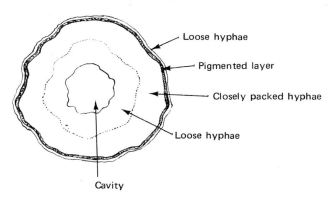

- Loose hyphae
- Pigmented layer
- Closely packed hyphae
- Loose hyphae
- Cavity

Plate 6.8   Sub-cortical rhizomorphs of *Armillaria mellea* as seen on removing the bark from a killed tree

6.5 NUTRITION

Fungi are unable to make use of solar energy because they lack a light-absorbing pigment such as chlorophyll. They need energy-rich substrates to supply their metabolic needs and are therefore described as HETEROTROPHIC in their mode of nutrition.

Because many substrates they use are insoluble they have first to make them soluble by means of digestive enzymes secreted through the tips of growing hyphae. As the actual process of digestion occurs outside the hyphae the enzymes are said to be extracellular. The types of chemical changes effected are many and varied but hydrolysis and oxidation are common.

The basic ingredients required for fungal growth and metabolism are:

1. An organic source of CARBON. Fungi can utilize almost every naturally occurring organic compound with various degrees of success but each species prefers a particular carbon source. Thus certain species of water moulds found in sewage grow only on fatty acids while most phycomycetous fungi can utilize only simple sugars such as glucose, fructose, mannose and maltose. Some mucoraceous moulds—for example, *Mucor mucedo*, which grows on bread—are able to digest starch. Polysaccharides such as cellulose are used mainly by members of the other three classes of fungi but these groups also include many sugar fungi such as yeasts living naturally on nectaries and on fruits. Lignin is broken down by only a few basidiomycete species such as bracket fungi which grow on trees; while other non-carbohydrate substrates used include amino-acids, sugar alcohols (for example, mannitol), organic acids and lipids.

These substrates provide the energy for anabolic (synthetic) reactions or act as raw materials for the building of new cell walls and organelles. Excesses are stored in the hyphae as glycogen granules and oil globules.

2. A source of NITROGEN. Fungi use simple inorganic nitrogenous compounds such as ammonium salts and nitrates, or organic substances—for example, amino-acids, amides and protein derivatives. As with carbon, they usually prefer a particular source of nitrogen. In common with all living organisms, fungi require nitrogenous compounds for the synthesis of proteins and the nitrogenous bases present in nucleic acids.

3. MAJOR MINERAL ELEMENTS which are required in substantial quantities include phosphorus as phosphate ions, sulphur as sulphate ions (although sulphur-containing amino-acids can form an alternative source of this element), potassium and magnesium. Some species also require calcium. Phosphate is necessary for the production of ATP, phospholipids essential for membrane structure, and the nucleic acids. Sulphate is needed to synthesize sulphur-containing amino-acids such as cystein and methionine which are present in many proteins. The role of the other elements is uncertain but they may activate specific enzymes or regulate membrane permeability.

4. TRACE ELEMENTS required in minute quantities include iron, zinc, copper, manganese and molybdenum. Some of these are thought to facilitate the formation of enzyme—substrate complexes thereby accelerating catalysis by enzymes, while others serve as electron carriers during redox reactions. Iron, for example, is present in cytochromes which accept electrons from substrates oxidized by hydrogen removal during respiration.

5. VITAMINS (organic accessory growth factors). Vitamins are usually involved with specific enzymic reactions such as the

synthesis of amino and fatty acids. Thiamin (vitamin B_1) is required by many species, while pyridoxine (vitamin B_6) and biotin (vitamin H) are needed in a few cases. Not all fungi are dependent on a supply of these compounds. Yeasts, for example, are noted for their production of vitamins and are used in the preparation of vitamin-rich extracts such as marmite for human consumption.

Environmentally, we can group fungi as saprophytes, parasites or symbionts:

(a) Saprophytes

Fungi which use only dead organic matter or organic substrates released from living organisms as sources of nutrients are usually called OBLIGATE SAPROPHYTES. Many are an important part of the microbial populations of soils where they bring about the recycling of nutrients and energy. Every year in a temperate deciduous forest between 500,000 and 1 million kg of organic matter is added to each km^2 of soil through leaf and branch fall and root production. In a tropical forest the figure is thirty to forty times this amount. Yet the amount of humus in undisturbed soils changes little from year to year. The equilibrium is maintained through the activities of soil saprophytes.

The organic substrates entering soils are extremely varied in nature. The simpler soluble materials such as sugars, amino-acids, other organic acids and inorganic minerals can be absorbed unchanged by soil fungi, bacteria and actinomycetes. Insoluble substances such as starch, cellulose, lignin, fats, oils, waxes and resins are made soluble by the action of enzymes secreted by soil micro-organisms. Cellulose is hydrolysed by the enzyme CELLULASE to cellobiose which is absorbed through the hyphal walls and then converted to glucose. Apart from fungi, few organisms can digest cellulose. Basidiomycete fungi are mainly responsible for the breakdown of lignin. Exactly how lignin is decomposed is not clear but it is broken down into a number of organic acids which a wide variety of fungi can utilize. Starch is hydrolysed by AMYLASES into the soluble sugar maltose, which, like cellobiose, is absorbed and then converted to glucose. LIPASE enzymes hydrolyse lipids such as fats and oils into glycerol and fatty acids. The peptide linkages in proteins and polypeptides are broken by PEPTIDASES to yield amino-acids. Some amino-acids may be absorbed unchanged, others are oxidized to release ammonia. Soil fungi therefore have a key role in the decomposition of organic matter.

It is usual to consider the colonization of organic substrates by fungi as a nutritional SUCCESSION. Primary sugar fungi appear first and utilize the soluble carbon sources; they are followed by cellulose- and then lignin-decomposers. Secondary sugar fungi often appear in the last two phases utilizing some of the excess soluble substrates released by the cellulolytic and ligninolytic species. A familiar example is the succession on herbivore dung (Table 6.1, Fig. 6.16).

Although this is called a nutritional succession, the sequence may not be as simple as it seems at first sight. The changes in fact merely reflect the time taken for the development of the fruiting bodies of the various groups of fungi. All the species are active from the start and compete for soluble carbonaceous compounds such as sugars. When the sugars have been used

Fig. 6.16   Coprophilous fungi from herbivore dung.

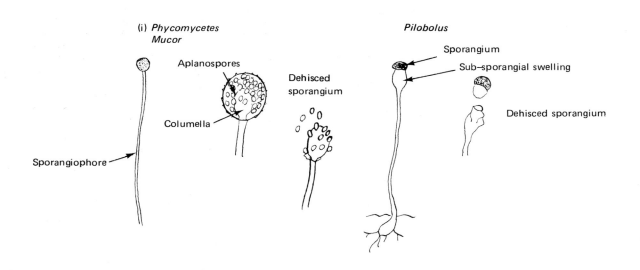

(i) *Phycomycetes*
Mucor

Aplanospores

Columella

Dehisced
sporangium

Sporangiophore

Pilobolus

Sporangium

Sub-sporangial swelling

Dehisced sporangium

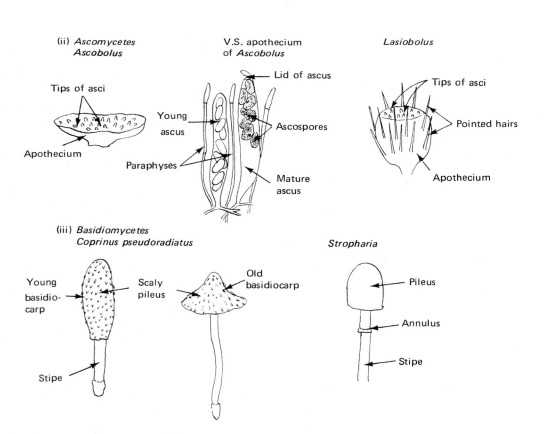

(ii) *Ascomycetes*
Ascobolus

Tips of asci

Apothecium

V.S. apothecium
of *Ascobolus*

Lid of ascus

Young
ascus

Ascospores

Paraphyses

Mature
ascus

Lasiobolus

Tips of asci

Pointed hairs

Apothecium

(iii) *Basidiomycetes*
Coprinus pseudoradiatus

Young
basidio-
carp

Scaly
pileus

Old
basidiocarp

Stipe

Stropharia

Pileus

Annulus

Stipe

up the ascomycete and basidiomycete fungi compete for cellulose by which time the phycomycete species will have sporulated. Eventually lignin is left which only the *Basidiomycetes* can degrade. Meanwhile the *Ascomycetes* will have fruited, leaving the basidiomycete fungi to reproduce last of all.

Comparable patterns of fungal succession have been reported on many natural substrates. The actual species which appear during the various phases will largely depend on the chemical composition of the material. On decomposing plant leaves the phycomycetous phase is absent, while on roots the primary colonizers are often imperfect fungi. Keratinolytic fungi dominate the final stages of decomposition of owl and hawk pellets which contain hair and feathers of the prey.

Table 6.1 Fungal succession on herbivore dung

FUNGI	APPEAR AFTER	LAST FOR	SUPPOSED ROLE
Phycomycetes e.g. *Mucor, Pilobolus*	1–3 days	7 days	Primary sugar fungi
Ascomycetes e.g. *Coprobia, Ascobolus*	5–6 days	3–4 weeks	Cellulolytic fungi
Basidiomycetes e.g. *Coprinus, Stropharia*	9–10 days	3–4 weeks	Ligninolytic fungi

(b) Parasites

Parasitic fungi inhabit the tissues of living organisms (hosts). Infections caused by ECTOPARASITES are confined to superficial tissues. ENDOPARASITES are responsible for deep-seated diseases. Those fungi, which grow only on or in living tissue and which cannot be easily cultivated on artificial media, are known as OBLIGATE PARASITES. An example is *Phytophthora infestans,* the cause of potato blight. Other parasites can live saprophytically at some stage and are called FACULTATIVE PARASITES. *Pythium debaryanum,* for example, causes the fatal condition known as damping-off when it parasitizes seedlings. It then decomposes the dead tissue of its host.

A host can be infected by hyphae growing actively through soil, by germinating spores which have been dormant in soil or by means of spores dispersed aerially from diseased plants. Spore germination is often stimulated by sugars or amino-acids secreted by the roots or leaves of host plants although many spores will produce germ tubes in the absence of stimulants. Unprotected surfaces such as the epidermis of young roots, which lack a cuticle, natural pores (stomata and lenticels) and damaged tissues are colonized by hyphae. Some pathogens such as *Erysiphe graminis* form special structures which penetrate the cuticle of stems and leaves (Fig. 6.17). Once inside the host most obligate parasites grow between the cells of parenchymatous tissue and push swollen, food-absorbing bulges known as HAUSTORIA into the host's cells.

Not all higher plants are susceptible to attack by pathogenic fungi and much interest has been focused on resistance mechanisms. In some instances the host is unable to produce a nutrient such as an essential vitamin. Compounds which prevent fungal growth are formed in other cases: prussic acid (HCN) is secreted by the roots of strains of flax plants which are resistant to wilt disease caused by *Fusarium oxysporum*. These are examples of PASSIVE resistance mechanisms. ACTIVE resistance occurs where plants produce a response only after infection. Many potential hosts produce chemicals called PHYTOALEXINS which provide immunity when they are attacked by pathogenic fungi. In other plants there is sudden local death of the host tissue when a pathogen enters, preventing the parasite from establishing a nutritional base. This is known as the HYPERSENSITIVE RESPONSE.

Fig. 6.17   Early stages in infection of a wheat leaf by *Erysiphe graminis*

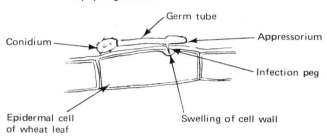

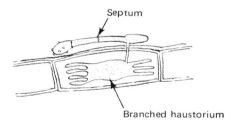

(c) Symbionts

SYMBIOSIS is the living together of two organisms to their mutual benefit. One or even both of the partners may be incapable of living alone, but other symbionts show varying degrees of interdependence. Two kinds of fungal symbiosis are (1) MYCORRHIZAE and (2) LICHENS.

(1) A mycorrhiza is a composite structure consisting of the root of a higher plant and a fungus. In ECTOTROPHIC MYCORRHIZAE the fungus forms a sheath which invests the lateral roots of many forest trees such as oak, beech, pine, larch and spruce (Plate 6.9). The fungal partner is often a basidiomycete (for example, *Boletus*), unable to hydrolyse lignin and cellulose, the most common carbon-rich substrates in soil. Experiments in which $^{14}CO_2$ was fed to mycorrhizal pine seedlings showed that some radioactive photosynthetic products were translocated to the fungal sheath, illustrating the dependence of the fungus on the higher plant for organic food. Surprisingly, despite the drain on their metabolic resources infected plants display a more rapid growth rate than those without the fungal partner (Table 6.2), possibly because of growth-promoting substances produced by the fungi.

Table 6.2 Comparison of the growth shown by mycorrhizal and non-mycorrhizal plants (composite data from Harley, 1971)

| PLANT SPECIES | AVERAGE DRY WEIGHT OF SEEDLINGS OR CUTTINGS (g) | |
	NON-MYCORRHIZAL	MYCORRHIZAL
Pinus strobus	0·303	0·405
Quercus robur	1·14	1·69
Eucalyptus macrorhiza	7·8	11·3

Plate 6.9 T.S. Ectotrophic mycorrhiza, x 100

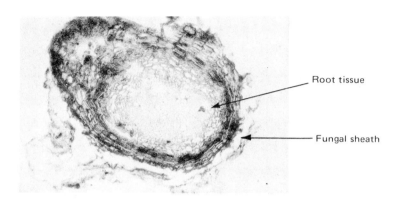

Root tissue

Fungal sheath

Other experiments have shown that the uptake of mineral nutrients, especially of nitrogen, phosphorus and potassium, is more rapid in mycorrhizal plants (Table 6.3).

Table 6.3 Uptake of N, P and K by Pinus strobus (after Hatch, 1937)

| | NUTRIENTS ABSORBED AS % DRY WEIGHT | | |
	N	P	K
Mycorrhizal	1·24	0·196	0·744
Non-mycorrhizal	0·85	0·074	0·425

In ENDOTROPHIC MYCORRHIZAE the amount of hyphae on the root surface is rather scant, most of the mycelium being inside the root and intracellular (Plate 6.10). Many orchids form endo-trophic mycorrhizae with basidiomycete or imperfect fungi. The relationship here is possibly one of controlled parasitism because the fungal partner is kept in check through periodical digestion of intra-cellular hyphae with enzymes secreted by the root cells. However, the fungus rarely causes death of its host, which is often dependent on the fungus for survival. Orchid seeds are not naturally infected with the fungus. At the start of germination they use their endospermic reserves. There then follows a phase when roots of the host are invaded by the fungus which hydrolyses soil cellulose and lignin and passes on the soluble absorbed products to the seedling, which is thus acting as a saprophyte. Some orchids such as *Neottia nidus-avis* (British bird's nest orchid) remain saprophytic throughout their lives. Others photosynthesize but may still remain partially dependent on the fungus for supplies of carbonaceous compounds. The photosynthetic—saprophytic balance varies from species to species.

Plate 6.10 T.S. Endotrophic mycorrhiza, x 200

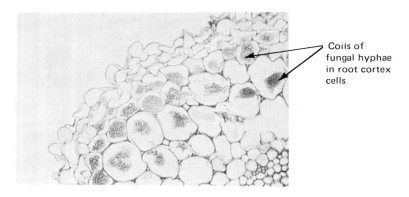

Coils of
fungal hyphae
in root cortex
cells

Plate 6.11   Lichen Morphology:

1. Crustose lichen—*Xanthoria* × 1

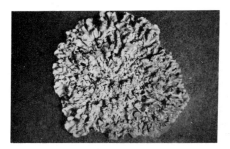

3. Fruticose lichen—*Usnea,* × 1

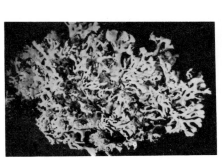

2. Fruticose (shrub-like) lichen
   — *Cladonia* x 1

(2) A lichen is an association between an alga (phycobiont) and a fungus (mycobiont). The fungus often makes up over 90% of the dry weight of the composite plant. About thirty

80

different genera of algae have been found in lichens, the commonest being the green, unicellular *Trebouxia* which interestingly has never been found free-living. The fungal partner, which never occurs as a free-living organism, is nearly always an ascomycete.

There are about 18,000 species of lichen. Many form crust-like growths on the surfaces of tree bark, stones, roofing tiles and so on; some are shrub-like, others leaf-like (Plate 6.11).

The anatomy of lichens is variable but they are often flat with the alga cells (gonidia) confined to a distinct layer near the upper surface (Plate 6.12).

Plate 6.12 Lichen anatomy:

1. V.S. *Parmelia* thallus, x 200

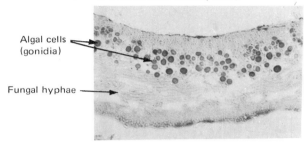

Algal cells (gonidia)

Fungal hyphae

3. V.S. Apothecium of *Xanthoria*, x 200

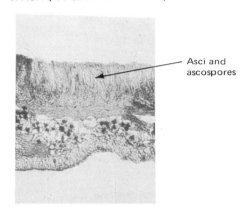

Asci and ascospores

2. Apothecia of *Xanthoria,* x 5

Both sexual and asexual methods of reproduction take place but not necessarily in the same species. Vegetative propagation may simply depend on fragmentation of the lichen thallus or may involve the production of characteristic reproductive bodies known as SOREDIA, each consisting of a few algal cells surrounded by fungal hyphae (Fig. 6.18). They are visible as fine, powdery masses on the surfaces of lichens. Under suitable conditions a soredium germinates to form a new thallus.

Fig. 6.18   Lichen soredia

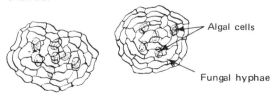

Algal cells

Fungal hyphae

The asci are often held in a saucer-shaped structure, the APOTHECIUM (Plate 6.12). The liberated spores must of course come into contact with the right alga in order to synthesize a new lichen thallus. Although the chances of this may seem remote lichens are long-lived so there is no need for a rapid replacement rate.

Lichens are extremely resistant to drought, nutrient shortage and extremes of temperature under which conditions they become dormant. Metabolic activity quickly resumes when water is available and the uptake of minerals is then very rapid. The efficiency with which they absorb solutes may account for the sensitivity of many lichens to atmospheric pollution (Fig. 6.19, Table 6.4).

Table 6.4 Comparison of the absorption of atmospheric sulphur by a lichen and inert substrates (after Gilbert, 1969)

MATERIAL	INCREASE IN S CONTENT OVER 46 DAYS (PPM)
Cotton wool	248
Glass wool	342
Living *Usnea*	1,010

Fig. 6.19   Distribution of lichens in the Newcastle-on-Tyne area (after Gilbert, 1970)

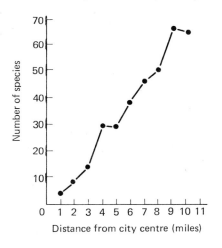

From the small number of studies that have been made on lichen physiology it appears that about half the photosynthetic carbohydrate made by the alga is absorbed into the fungus. Where the phycobionts are blue–green algae atmospheric nitrogen can be reduced to ammonia, most of which is passed on to the mycobiont. It is difficult to see what benefit the alga gains from the lichen association because there is little or no evidence for the passage of any substance from the mycobiont to the phycobiont. However, it is probable that many more types of habitats are open to the alga when it lives in the symbiotic state.

7 GROWTH OF MICRO-ORGANISMS

The word growth usually implies an increase in size but a change in size is not the only criterion for measuring growth in living organisms. For example, when a fungal spore puts out a germ tube it is said to be growing. While there can be no doubt that it is increasing in size it is also losing dry weight because stored nutrients are used to provide the energy necessary for germination. However, when the fungal colony is established on a nutrient medium its volume and dry weight gradually increase in time. We may thus define growth as an orderly and permanent change in any measurable property such as volume, dry mass, nitrogen content and so on. The rate of growth is the amount of change in a given period of time.

7.1 UNICELLULAR MICRO-ORGANISMS

It is not easy to examine the growth patterns of individual unicellular microbes because of their smallness. Instead it is simpler to study the growth of populations, in which case growth refers to the increase in the number of individuals.

Bacteria and yeasts multiply in such a way that the number of organisms usually doubles at each generation (Fig. 7.1). As some of these organisms produce a new generation every twenty to thirty minutes, their numbers soon become astronomical. Of course, not all the progeny survive so the TOTAL count, dead and living, is usually different from the VIABLE count. Nevertheless, the earth is not overrun with micro-organisms even though they can grow almost anywhere. It is obvious, therefore, that the rapid growth of populations of microbes is soon checked in nature.

Fig. 7.1   Growth of a population of bacteria

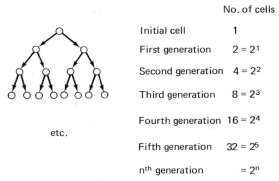

	No. of cells
Initial cell	1
First generation	$2 = 2^1$
Second generation	$4 = 2^2$
Third generation	$8 = 2^3$
Fourth generation	$16 = 2^4$
Fifth generation	$32 = 2^5$
n^{th} generation	$= 2^n$

etc.

If we inoculate a liquid culture medium with bacteria or yeast cells and count the numbers of cells present at different intervals of time we can construct a GROWTH CURVE. Try this for yourself from the data in Table 7.1. Because the numbers involved are so large it is easier to plot the log of the number of cells against time. You will then obtain a curve as shown in Fig. 7.2. The curve can be divided into a number of distinct phases, the length of each phase varying from species to species and the conditions under which growth is taking place.

Table 7.1 Growth of a population of Escherichia coli *in nutrient broth*

HOURS	NO. VIABLE CELLS/CM³ MEDIUM	TOTAL NO. CELLS/CM³ MEDIUM
0	20,000	20,000
2	21,900	27,200
4	496,000	540,000
6	5,430,000	6,400,000
8	81,900,000	105,760,000
12	83,400,000	126,300,000
24	80,500,000	127,600,000
36	1,120,000	127,900,000

Fig. 7.2   Growth curves for a population of *Escherichia coli*

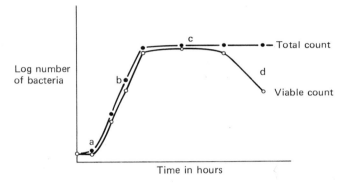

(a) The lag phase

During the lag phase there is little or no increase in cell numbers although each bacterium may increase in size. The length of this phase depends largely on the nature of the inoculum, the medium to which it is transferred and conditions of incubation such as temperature. If the bacteria in the inoculum are in an active state then they will multiply almost immediately but if they are dormant they may not divide for some time. Should the composition of the medium be different from that on which the inoculum was grown then the organism will require some time to adapt to its new diet. As growth is an enzyme controlled process it follows that the temperature must also be suitable for enzyme action.

(b) The log phase

Here the logarithm of the cell number increases linearly with time. This part of the growth cycle is often called the EXPONENTIAL phase. An exponent is an index number written above another indicating the power of the latter; for example, $a^2 = a \times a$.

84

If we begin with 20,000 bacteria/cm^3, we will have:

after one generation $\quad 20{,}000 \times 2^1 = 40{,}000$/cm^3
after two generations $\quad 20{,}000 \times 2^2 = 80{,}000$/cm^3
after n generations $\quad 20{,}000 \times 2^n$.

Another way of stating this relationship is:

$$N = N_0 \times 2^n$$

where N_0 = original number of cells and N = number of cells after n generations,

or $\quad \log N = \log (N_0 \times 2^n) = \log N_0 + n \log 2$

This equation can be used to calculate the numbers of cells expected at any time during the log phase, the generation time and the number of generations required to produce a certain number of cells. For example, Fig. 7.2 indicates that the exponential phase began after two hours at which time there were 21,900 bacteria/cm^3 and after four hours there were 496,000 bacteria/cm^3.

Thus $N_0 = 21{,}900$ *2 hrs.*
and $N = 496{,}000$ *4 hrs.*

We can now calculate n as follows:

$$496{,}000 = 21{,}900 \times 2^n$$

or $\quad \log 496{,}000 = \log 21{,}900 + n \log 2$

$$\therefore n = \log \frac{496{,}000 - \log 21{,}900}{\log 2}$$

$$n = \frac{5 \cdot 6955 - 4 \cdot 3404}{0 \cdot 3010}$$

$$= \quad 4 \cdot 5.$$

In two hours, or 120 minutes, we have produced 4·5 generations. The generation time, time to produce one generation, is therefore

$$\frac{120}{4 \cdot 5} = 26 \text{ minutes.}$$

Now calculate for yourself the number of bacteria per cm^3 after ten hours in 20 cm^3 of a medium inoculated with 5,000 cells assuming that growth was exponential from two hours onwards and that the organism had a generation time of twenty minutes.

(c) The stationary phase

During the stationary phase the population number remains static and the birth rate is almost balanced by the death rate. In a batch culture such as we have grown, one of the most potent growth-limiting factors will be the amount of available food because the culture vessel contains a finite quantity of medium. Just as the sizes of human populations are checked by starvation, so with bacteria. When the energy content of available substrates is less than the energy required by the population, a check in growth rate is inevitable. Apart from a depletion in nutrients, changes in the environment may be made by microbes themselves. Metabolic products may become toxic as they accumulate; for example, yeasts produce ethanol by fermenting grape juice but when the alcohol content reaches about 15% the yeasts are killed. Other microbes change the pH of their surroundings making it unsuitable for continued growth; for example, *Lactobacillus* which produces lactic acid from lactose during the souring of milk.

(d) The death phase

During the death phase those individuals unable to compete successfully will die. There is now a marked difference between the total and viable counts. As cells die they undergo a process of self-digestion or AUTOLYSIS. Enzymes which previously were carefully regulated by the organisms now wreak havoc and digest the cell structure. Some cells still retain their structural and physiological integrity and feed on the autolytic products of others. Ultimately the amount of energy-rich material available for microbial growth becomes nil and no organism can then survive in an active state. Spore-producing bacteria may be able to survive in a dormant condition.

7.2 FILAMENTOUS ORGANISMS

(a) Growth on solid media

Fungi grow by hyphal elongation, thus the rate at which a fungal colony increases in diameter on a solid medium is a useful measure of fungal growth. The method must be used with some caution because the density of hyphae often varies considerably on different media. Nevertheless colony diameter is a fairly reliable comparative guide when used to investigate the effects of environmental factors such as temperature on growth rates of fungi using a defined medium.

Plot a graph of the figures in Table 7.2 using colony diameter as the vertical axis and time as the horizontal axis.

Table 7.2 Growth of a colony of Trichoderma *on an agar medium*

TIME IN DAYS	1	2	3	4	5	6
Colony diam. (mm)	6·0	21·0	37·0	53·0	62·0	71·0

Three phases are seen in the growth pattern:

1. A lag phase. Spores rarely germinate immediately they are inoculated on to a fresh medium. If a mycelial disc is used as the inoculum there will be many broken hyphae which must be repaired before growth commences. In either case the inoculum may have to adapt to new conditions of nutrient supply, aeration, temperature and so on. When the inoculum is old the lag phase is usually long.

2. A phase of rapid growth. At this stage the colony diameter extends linearly with time; there is a constant increase in size for a given interval of time.

3. A phase of retarded growth. The growth of hyphae slows down, usually because of the accumulation of toxic wastes, staling substances, which may be accompanied in some species by autolysis of the hyphae at the centre of a colony. Reduced growth may also occur in response to the proximity of a physical barrier such as the edge of a petri dish.

(b) Growth in liquid media

Mycelial mats and pellets can easily be harvested by filtration and their dry masses determined. Increases in dry mass show the various phases of growth quite clearly (Fig. 7.3).

1. A lag phase. Can you suggest reasons for this period of no growth?

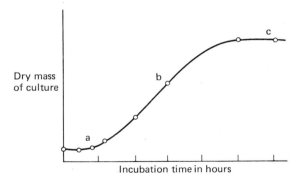

Fig. 7.3 Growth curve for *Mucor hiemalis* (after Trinci, 1972)

Dry mass of culture

Incubation time in hours

2. An accelerated growth phase in which the cube root of the dry weight increases linearly with time for some species. In *M. hiemalis* however the dry mass is doubled every four hours so that for this fungus this is a phase of exponential growth. The doubling time is as little as one hour in other species.

3. A stationary phase which may be due to the same factors as in stationary phases mentioned earlier.

4. A death phase occurs if liquid cultures are maintained for long periods of time. The dry mass decreases owing to autolysis of the mycelium.

7.3 FACTORS AFFECTING MICROBIAL GROWTH

Microbes will grow actively only if all external conditions are suitable.

(a) Temperature

Every living organism has a minimum temperature below which it cannot grow or dies, a maximum temperature above which it dies and an optimum temperature at which it grows most successfully. Among the bacteria and fungi there occur:

(*i*) PSYCHROPHILIC species which grow actively between 0 and 25°C, with an optimum temperature of below 20°C. Many are responsible for the deterioration of refrigerated food, others grow in low temperature soils in polar areas.

(*ii*) MESOPHILIC species which grow between 20 and 45°C, with an optimum temperature of 25–37°C. Most bacteria and fungi belong to this category, including the mammalian pathogens.

(*iii*) THERMOPHILIC species which generally grow at 45–60°C, with an optimum of 50–55°C. Fungi and bacteria which are active in the heating of manures and composts are thermophiles. The bacterium *Thermus aquaticus*, found in hot springs, amazingly grows at a temperature of 79°C.

Even though active growth of many microbes does not occur at very high or at very low temperatures they are able to survive these extremes by spore formation or by remaining in a dormant state. Thus in order to kill some microbes it is necessary to autoclave them at temperatures well above boiling point. Freezing foods simply prevents microbial growth without necessarily killing the organism present. In a domestic refrigerator the temperature drops to only 5°C, allowing the slow growth of microbes and gradual spoilage of food. In a freezer, at −18°C, most organisms are dormant.

(b) pH

Fungi are acid-tolerant and will grow at a pH as low as 3·5 although their optimum range is pH 5–6. They are therefore of considerable importance in the decay of organic matter in the acid soils of heaths and moorlands. Bacteria generally do not tolerate acid conditions and grow best at a pH of 6·5–7·5. However, there are exceptions. Certain species of bacteria such as *Thiobacillus thiooxidans*, which oxidizes sulphur to sulphuric acid, can grow at a pH of 0, while urea-hydrolysing species isolated from urinary tract infections can raise the pH of their environment to about 11 by releasing ammonia. Because microbes readily alter the pH of their surroundings, media are generally buffered with phosphate salts.

(c) Oxygen

Reference has previously been made to the effect of oxygen on bacterial growth (Chapter 2). Much the same applies to the growth of fungi, algae and protozoa.

Growth is an ENDOTHERMIC process and energy is produced in all living organisms by the oxidation of energy-rich substances such as sugars. In the case of aerobes the function of oxygen is to act as the final acceptor of hydrogen removed from substrates which have been oxidized. The hydrogen is transferred to oxygen by hydrogen-acceptor molecules such as co-enzymes I (NAD) and II (NADP), flavoprotein (FADN) and cytochromes:

Dehydrogenase enzymes catalyse the reduction of the hydrogen acceptors. The oxidation of reduced cytochrome is catalysed by the enzyme cytochrome oxidase.

At a number of points in this respiratory chain energy is liberated and stored in the high energy bonds of ATP. In the absence of oxygen however, the respiratory chains are unable to operate, substrates are only partially oxidized, and much less energy is released. The anaerobic oxidation of a mole of glucose by a yeast, for example, yields only two high energy phosphate bonds compared with 38 formed by aerobic respiration. Clearly, the anaerobic method of energy release is highly inefficient and can only be indulged in by organisms which live in an environment where there is abundant sugar. Small wonder that most microbes which rarely experience such a luxury utilize the aerobic mechanism.

(d) Nutritional factors

The nutritional requirements of micro-organisms are extremely diverse. They are discussed in Chapters 2, 3, 4, 5 and 6. To summarize:

(*i*) Heterotrophs require a source of carbon, usually organic; nitrogen, organic or inorganic nitrogenous substances; minerals and vitamins.

(*ii*) Photoautotrophs require light, carbon dioxide and a source of hydrogen to reduce the products of CO_2 fixation.

88

Blue—green and green algae employ water for the hydrogen supply but in photoautotrophic bacteria H_2S may be used instead.

(*iii*) Chemoautotrophs have much the same requirements as photoautotrophs but do not require light.

Growth of a micro-organism may be limited by one or more of its dietary needs. For each factor there is an optimum value which supports maximum growth. When minerals, vitamins and amino-acids are at low concentrations there is often a linear relationship between the quantity of available substrate and the amount of growth (Fig. 7.4). This linear relationship can be used in assaying the quantities of available vitamins in foodstuffs, blood and so on.

Fig. 7.4 Linear relationship between growth of *Lactobacillus arabinosus* and amount of niacin in medium.

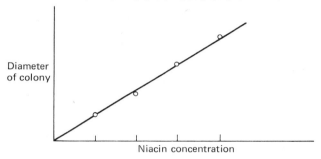

There are occasions when the concentration of nutrients may be too high for microbes. The most obvious practical application of this is the preparation of foods of high osmotic potential such as jams and honey. However, a few micro-organisms can adapt even to such extreme habitats. These OSMOPHILIC species include certain yeasts which can live on substrates containing as much as 70—80% sugar.

Microbial growth in most ecological habitats is restricted by the lack of available nutrients as well as by other factors such as aeration, temperature and pH. There are times, however, when raising nutrient levels causes the growth of micro-organisms to flourish to the detriment of other organisms in certain habitats. Thus the discharge of raw sewage into streams and rivers results in the proliferation of aerobic microbes which quickly reduce the oxygen level of the water, making it unsuitable for fish and other animals. The leaching of nitrates and phosphates into streams and rivers from arable soils to which fertilizers have been applied stimulates the formation of algal blooms especially of blue—green algae such as *Microcystis* and *Anabaena*. The subsequent decay of the algae by aquatic microbes creates anaerobic conditions (see Chapter 8, Section 5) and can also result in the release of toxins which poison fish and other animals.

(e) Antibiotics

The micro-organisms that are capable of active growth in any given habitat must be able to utilize the available substrates, tolerate the prevailing climatic conditions and survive the intense competition from other organisms. Some display an additional asset; they produce substances which are toxic to other forms of life. Among such toxic substances can be included the antibiotics, a group of naturally synthesized chemicals which are specific in their inhibition of microbial life even at very low concentrations. Since the discovery of penicillin by Fleming in 1929, a wide array of antibiotics has been isolated. Some are of fungal and bacterial origin but the majority come from the actinomycetes. Interestingly, many of the producers are soil-dwelling organisms and there is some evidence that they prevent the growth of antibiotic-sensitive micro-organisms in such habitats. Because other factors such as growth rate are also of significance in the colonization of substrates, many non-antibiotic producing organisms are successful in the soil ecosystem. Other microbes are able to produce enzymes which destroy the antibiotic. Penicillinase secreted by some species of bacteria renders penicillin useless against them.

In pure culture some antibiotics such as penicillin inhibit bacterial growth by preventing the synthesis of wall mucopeptides. Naked bacterial protoplasts, formed at cell division, then absorb water osmotically and undergo lysis. In other cases, as with chloramphenicol, inactivation of RNA occurs so that protein synthesis, and hence growth, is prevented. A few antibiotics, griseofulvin for example, are effective against fungi, but none seem to affect viruses, so that antibiotics cannot be used against viral diseases.

(f) Other factors

Ultra-violet light is frequently used to sterilize air and equipment. This and other forms of short-wavelength electromagnetic radiations such as X-rays kill microbes by damaging the nucleic acids. Thus the normal process of replication does not occur and growth is inhibited.

8 MICRO-ORGANISMS AND MAN

For many centuries man lived alongside micro-organisms without any knowledge of the relationship between them and his own health and survival. During this time however he unwittingly made use of some microbial metabolic processes in the preparation and production of foods such as bread and cheese, drinks such as wines and ales, and textiles such as the retting of flax. Without knowing what was happening or even the fact that micro-organisms were involved at all, many of these processes were controlled and perfected to a relatively high scientific standard.

The publication in 1857 by Louis Pasteur of a study of wine fermentation established the beginnings of the science of microbiology. The acceptance by the medical profession of the role of pathogenic micro-organisms in human disease together with Pasteur's work on the chemistry of fermentations led to a rapid improvement in knowledge of micro-organisms and their activities. In the short time since Pasteur we have not only learned how microbiological-based processes work but also how to improve traditional methods and develop entirely new processes to our own advantage. The use of micro-organisms by man is termed industrial microbiology.

8.1 MICROBIOLOGICAL FERMENTATIONS

(a) Production of alcohol

The fermentation of plant materials by yeast to produce ethanol was the earliest use man made of micro-organisms. As long ago as 3000 BC the Ancient Egyptians could brew alcohol. Today however considerable quantities of alcohol are used in the manufacture of such commodities as paints, plastics and textiles, but most of the ethanol produced at the present time is consumed in the form of alcoholic drinks. Two of the principal fermented beverages are wine and beer.

1. Wine production. Ripe grapes are picked and crushed to produce a MUST which contains juice, pulp, skins and pips. The must is then treated with about 100 p.p.m. of sulphur dioxide to inhibit undesirable bacteria and yeasts present on the skins. Fermentation of the juice is brought about by wine yeast, *Saccharomyces cerevisiae var. ellipsoideus,* which also grows on the skins of the fruit (Plate 8.1). Some wine producers add starter cultures of known strains of the fungus. As fermentation proceeds alcohol is formed which extracts the red pigments located in the skins of red grapes. If the skins are removed before the process is too advanced a pink or rosé wine is produced. White wine is produced if the skins are removed early or if the must is from green grapes. Complete breakdown of the sugar in the must produces a dry wine with an alcohol content of up to 15% at which concentration the yeast is killed. Sweet wines are made from musts with very high sugar content or by stopping fermentation before all the sugar is used up. The fermented liquor is then aged in vats, when sediment is deposited, then filtered before it is bottled.

Plate 8.1   Yeast bloom on grape skins (from Nuffield Advanced Biology Trial Scheme B.P.4)

Sparkling wines such as Champagne are made by adding sugar to the fermented must. Wine yeast is then added and the mixture bottled. Champagne bottle corks have to be held in place with wire to withstand the pressure of the carbon dioxide formed inside the bottle by further fermentation.

The characteristics of wine depend not only on the way in which the must is treated but also on the composition of grape juice. This will depend on the type of grape and the environmental conditions under which the fruit is grown. Each major wine-producing region uses the variety of vine most suited to its climate and soil.

2. *Beer.* Its production depends on yeast fermentation of an infusion made from grain, mostly barley, and compared to wine production is microbiologically a more controlled process.

The barley grains are germinated under carefully controlled conditions so that grain enzymes, amylases and peptidases, convert endospermic starch to maltose and protein to amino acids. The MALTING is brought to an end by raising the temperature to kill the barley embryos. The addition of warm water to the malt forms a liquor, the WORT. Dried hops are then added to give flavour, and to release chemicals which have antimicrobial properties. The wort is boiled for several hours.

Following cooling and filtering, the wort is PITCHED, that is inoculated with brewers' yeast. *Saccharomyces cerevisiae* is used for most beers but if a light beer or lager is being produced *S. carlsbergensis* is added instead. In the large fermentation vats the yeast converts soluble sugars in the wort into ethanol and carbon dioxide over a period of two to five days. A thick yeast head forms on the surface of the vat, and is periodically skimmed off. When the beer has reached the desired alcohol content, between 4–8%, it is racked off into casks or storage tanks, where a continued, secondary fermentation may occur at a slow rate. The beer is clarified, frequently enriched with carbon dioxide and is then ready for the market

92

(Fig. 8.1, Plate 8.2). INDUSTRIAL ALCOHOL is used as a solvent in the chemical industry. Some of the requirement is met by alcohol of microbiological origin. Any waste substance rich in carbohydrate, such as molasses from sugar refining or spent brewery grains, can be fermented with selected strains of *S. cerevisiae*. The demand for industrial alcohol is so great, however, that the catalytic hydration of ethylene using sulphuric acid is now a more rapid and economic process for the production of ethanol.

Fig. 8.1 Production of beer in a typical brewery

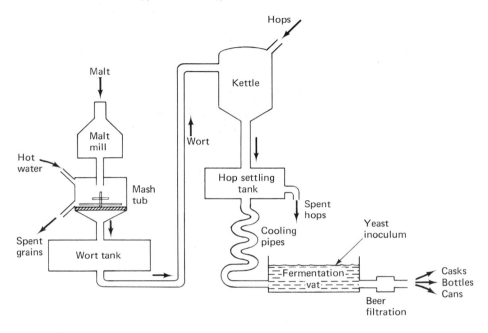

Plate 8.2   Production of beer

1.   Hop kettle for the preparation of wort

2.   Open tank fermentation – note yeast floating on surface of the wort

93

(b) Vinegar and acetic acid   Vinegar results from partial oxidation of ethanol to acetic acid:

$$(CH_2O)_n \longrightarrow CH_3CH_2OH \xrightarrow{O_2} CH_3COOH + H_2O$$

carbohydrate   ethanol   acetic acid

This souring of alcohol is a natural process brought about by the aerobic bacteria, *Acetobacter* and *Acetomonas*. The traditional Orleans process is widely used today (Fig. 8.2). Poor quality wine is held in partly filled wooden vats or casks and a starter bacterial culture added. The microbial oxidation is a slow process taking several weeks to complete but the product is of high quality. Vinegar is tapped off from the bottom of the casks and fresh wine added to restore the original volume.

Fig. 8.2 The Orleans vinegar process

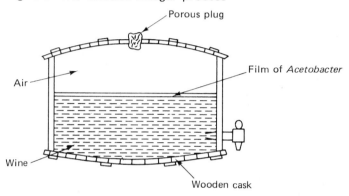

In Britain, most vinegar comes from malt wort or cider and a quicker process of manufacture is employed. Wort is allowed to trickle down tall towers packed with wood shavings which produce a large surface area for growth of *Acetobacter*. Oxidation of the alcohol occurs over a period of four to five days. Vinegar collects at the base of the tower, is tapped off and barrelled or bottled.

(c) Antibiotics   An antibiotic is a chemical substance produced by a micro-organism which has the capacity to inhibit or even destroy other micro-organisms in dilute solutions. Observations of the effects of the mould *Penicillium notatum* on certain pathogenic bacteria by Alexander Fleming in 1929 led to the isolation of penicillin. Demand for antibiotics on a large scale began at the start of the second World War, in 1940. The growth of the antibiotics industry during the following fifteen years shows how rapidly empirical discoveries linked with technology can become an applied science in microbiology. Most antibiotics are bacterio-static or fungi-static; they prevent growth of micro-organisms rather than kill them. They are selective in their action and whilst some such as the tetracyclines, streptomycin for example, are effective against several organisms (broad spectrum antibiotics), others such as penicillin act on one or only a few pathogens (narrow spectrum antibiotics). Their extensive use in medicine has enabled man to conquer many bacterial diseases (Table 8.1).

94

Table 8.1 Some antibiotics and the diseases they control

ANTIBIOTIC	PRODUCER	DISEASE	PATHOGEN
Penicillin (N.S.)	Mould: *Penicillium*	Pneumonia Diphtheria Tetanus Scarlet fever Skin infections	*Pneumococcus* *Corynebacterium* *Clostridium* *Streptococcus* *Staphylococcus*
Chloramphenicol* (B.S.)	Actinomycete: *Streptomyces*	Skin infections Typhoid fever Syphilis Gonorrhoea	*Staphylococcus* *Salmonella* *Treponema* *Neisseria*
Streptomycin (B.S.)	Actinomycete: *Streptomyces*	Tuberculosis Pneumonia	*Mycobacterium* *Pneumococcus*
Griseofulvin (N.S.)	Mould: *Penicillium*	Ringworm	*Trichophyton*

N.S.—Narrow spectrum.   B.S.—Broad spectrum.
* Now prepared by chemical synthesis.

Industrial production. Antibiotics are obtained commercially by culturing the producer micro-organism in a sterile culture medium. The scale of the fermentation process is vast, tanks with capacities of tens of thousands of gallons being employed. With agitation and aeration and under optimum temperature conditions, the micro-organism grows, secreting antibiotic products into the medium. Using chemical and biological assay techniques, the time when the antibiotic is at its highest concentration is detected. The organisms are then filtered out of the spent medium and the antibiotic is extracted by use of solvents or by precipitation (Fig. 8.3, Plate 8.3).

Plate 8.3 Antibiotic production.

1. Early trials with shaker flasks

Plate 8.3 (Continued)
2. A battery of 10,000 gallon antibiotic fermentation tanks

Penicillin. This is a general term used to describe a class of substances of related structure produced by the mould *Penicillium.* The chemical structure is complex and there are many different types of penicillin. Penicillin G, benzylpenicillin, is widely used clinically against Gram positive bacterial infections. Mutant strains of *P. notatum* and *P. chrysogenum,* produced artificially by ultra-violet irradiation, give high yields of the antibiotic.

Resistance to penicillin has been observed especially among the staphylococci which produce the enzyme penicillinase. This has led to the continued research for more effective penicillin derivatives. Many in use today are semi-synthetic products such as ampicillin. Penicillin acts by interfering with the formation of mucopeptide in cell walls of sensitive bacteria. Such cells become naked and the fragile protoplasts quickly rupture following the absorption of water.

Streptomycin. This is a particularly useful antibiotic of the broad spectrum type, and is used against both Gram positive and Gram negative pathogens and also to treat tuberculosis. It is produced by the actinomycete *Streptomyces griseus.* Other uses of streptomycin and its recent derivatives, such as novobiocin, lie in the treatment of bacterial plant diseases and also animal diseases caused by bacteria resistant to penicillin.

Its mode of action is not fully established but it appears to inhibit respiratory pathways in bacteria.

Chloramphenicol. This is a broad spectrum antibiotic originally isolated from the actinomycete *Streptomyces venezuelae*, but now prepared more economically by synthetic methods. It has proved to be very useful in the treatment of typhoid fever and venereal diseases and its mode of action is to interfere with bacterial protein synthesis.

Fig. 8.3   Flow chart for industrial production of an antibiotic

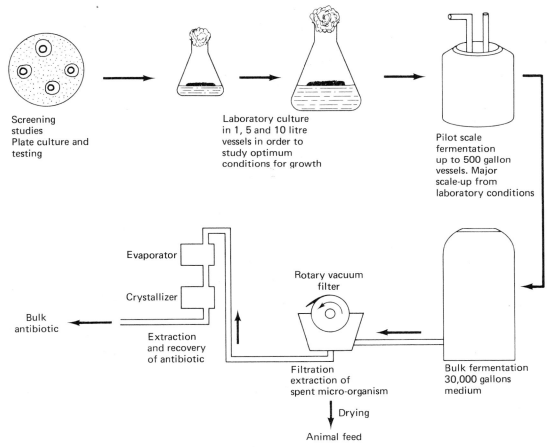

Screening studies
Plate culture and testing

Laboratory culture in 1, 5 and 10 litre vessels in order to study optimum conditions for growth

Pilot scale fermentation up to 500 gallon vessels. Major scale-up from laboratory conditions

Evaporator

Crystallizer

Rotary vacuum filter

Bulk antibiotic

Extraction and recovery of antibiotic

Filtration extraction of spent micro-organism

Bulk fermentation 30,000 gallons medium

Drying

Animal feed

8.2 FOOD PRODUCTION AND PROCESSING

The manufacture of fermented milk products such as butter, cheese and yoghurt, like brewing, traditionally used micro-organisms in a relatively uncontrolled way. The microbes concerned are generally termed lactic acid bacteria. They include bacteria naturally present in milk and pure strains which are grown in pasteurized milk and added as starter cultures.

(a) Butter

Butter is prepared by churning cream, causing the fat globules to coalesce into granules. The liquid portion, buttermilk, is then drained off and butter granules are compressed. Today, virtually all butter is made from pasteurized milk in which the natural microflora is killed but the cream is sometimes soured by starter cultures of *Streptococcus lactis* and *Leuconostoc citrovorum*. The streptococci convert milk sugar lactose to lactic acid while the leuconostocs produce volatile substances such as diacetyl which give the butter flavour and aroma.

(b) Cheese

In industrial cheese manufacture pasteurized milk is always used. Starter cultures of *Streptococcus lactis* or *S. cremoris* are added to the milk and changes in pH and texture are monitored as CURDLING proceeds. Rennet, a preparation from calves' stomachs containing the enzyme rennin, is usually added to speed up the coagulation of milk protein, casein. When soft cheeses are made the liquid portion or whey is allowed to drain from the curd but for hard cheeses the curd is compressed (Fig. 8.4).

Fig. 8.4 The manufacture of cheese.

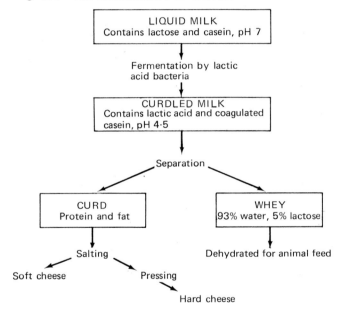

Sodium chloride is now mixed with the curd or applied to its surface. SALTING prevents the growth of undesirable microbes and adds flavour.

Unripened cheeses such as cottage cheese are nothing more than fresh, salted curd. However, most varieties of cheese are RIPENED by further microbial activity in or on the curd. In making blue cheeses the curd is inoculated with spores of *Penicillium roqueforti* and incubated in a humid atmosphere at 9–12°C. Lipase enzymes secreted by the mould hydrolyse fat, releasing fatty acids which contribute to the flavour of the cheese. The ripening of Camembert cheese is brought about by *P. camemberti* growing on its surface. This fungus produces peptidase enzymes which converts casein to compounds such as polypeptides, amino-acids and ammonia. Hard cheeses such as Cheddar are ripened by lactic acid bacteria growing in the curd. Although the bacteria do not secrete peptidases, small amounts of the enzymes are released by autolysis when the bacteria die, thereby causing some casein decomposition.

(c) Yoghurt

Yoghurt is one of several types of cultured buttermilks, originally prepared from the liquid removed from cream during the making of butter. Today it is made from pasteurized skimmed milk. Whilst popular in Asia and the Balkans for centuries, where it is produced from fermented goats' or mares' milk, it is a fairly recent introduction to the United Kingdom, but is now a most popular dairy food. In yoghurt production a mixed culture of *Streptococcus thermophilus* and *Lactobacillus bulgaricus* ferment the lactose in milk at temperatures of 42–45°C. The soured milk is somewhat unpalatable but its flavour may be improved by the addition of fruit pulps and juices.

8.3 MICROBES AS FOOD

Industrial fermentations have shown that the growth and metabolism of many micro-organisms can be controlled so as to produce useful products for man. The mass culture of micro-organisms for use as a potential food source is, however, relatively recent. Bacteria, yeasts, fungi and algae can all be grown to produce a crop. After washing, and drying, the resulting powder is nutritionally rich in protein, carbohydrate and fat. The production of low cost protein food on a vast scale is one way to help meet the need for additional food supplies for an increasing world population. The industrial microbiologist may well be the farmer of the future.

Micro-organisms may contain between 35 and 75% cell dry mass as protein which compares very favourably with conventional protein sources (Table 8.2).

Table 8.2 Protein levels of micro-algae and conventional sources

SOURCE	DRY MASS PROTEIN (kg/km^2/annum)
Spirulina sp. (blue-green alga)	24,310,000
Chlorella sp. (green alga)	15,600,000
Clover leaf	1,681,000
Grass	672,200
Fish	627,300
Peas	395,500
Milk	10,000

Protein from bacteria or yeast cells is of greater value to human diets than algal protein since it contains a greater variety of essential amino-acids. Further, bacteria and yeasts can be grown in continuous culture methods which are not possible for algal culture on the same scale. Suitable energy-rich substrates which are used to culture the cells include higher paraffins, methane and other hydrocarbons. In 1972 the production of such proteins by one of the major oil companies reached over 16,000,000 kg. Getting such unusual foods accepted by people is more likely to be a greater obstacle than the actual production.

99

8.4 THE DETERIORATION AND PRESERVATION OF FOOD

(a) Deterioration

The fitness or unfitness of food for human consumption depends very much on the judgement of the person about to eat that food. What one person will eat, possibly with relish, another will not. No housewife will willingly serve sour milk to her family yet will produce yoghurt for a dessert. She would be surprised if the family complained about the yoghurt being unfit to eat. Yet both sour milk and the yoghurt are essentially identical in containing very large numbers of lactic acid bacteria. Putrefying meat means meat undoubtedly spoiled and unfit to eat, yet putrefactive changes in Camembert cheese are normal to the ripening process and many people will pay for such a delicacy.

We take it for granted that food will always be available, reasonably priced, pre-packed and, above all, unspoilt, by which we really mean microbiologically sound. The food microbiologist must therefore be concerned with all stages in food production, from growing and throughout the collecting, preparing, packaging, distribution and the storing stages. At any one of these steps, deterioration may occur.

Some processes leading to food spoilage are not of microbial origin. Many vegetables and fruits deteriorate naturally as enzymes inside the cells break down the protoplasm. Such autolysis produces soft or brown foods spoiled in appearance and flavour. However, most spoilage is due to the activity of bacteria, moulds or yeasts. The smell, taste, colour or texture of the food may be sufficiently altered to make the food inedible.

A variety of factors determines the likelihood of spoilage of a particular food. These include its chemical composition, moisture content, acidity, temperature and oxygen availability, which together with the number and kinds of microbes will determine the rate and nature of decomposition.

Carbohydrate-rich foods are often spoiled by fungi. Moulds such as *Mucor, Rhizopus* and *Penicillium* soon grow on bread and cakes if they are kept in humid conditions. Inadequately dried grain quickly becomes mouldy during storage as a result of the growth of species of *Penicillium, Aspergillus* and *Fusarium*. Fresh fruit and vegetables, especially if damaged during harvesting or in transit, are rotted by fungi and bacteria. Soft rot of oranges caused by *Penicillium*, brown rot of apples due to *Monilia* and grey mould of strawberries attacked by *Botrytis* are familiar examples. Osmophilic yeasts and moulds sometimes ferment foods with a high sugar content such as honey and jam. In all of these forms of spoilage the food is made unpalatable, while some fungi especially those growing on moist grain and nuts produce AFLATOXINS which can cause death in man and farm animals if the mouldy food is eaten.

Foods with a high protein content are more commonly decomposed by bacteria. The smell of rotten eggs is due to the formation of hydrogen sulphide by protein-digesting species of *Pseudomonas*. Fresh and cooked meat and fish can be putrefied by a number of proteolytic bacteria including *Achromobacter* and *Pseudomonas*.

Untreated milk is soon soured by lactic acid bacteria (section 8.2, this chapter). The production of fatty acids such as butyric acid from milk fat by the action of lipolytic bacteria and fungi is the chief cause of milk becoming rancid. Ropy milk owes its name to the formation of long thin masses of slime by the bacterium *Alcaligenes*. Other bacteria occasionally found in milk secrete pigments which give the milk a pink or violet tinge. Cheese and butter may also develop coloured spots due to the growth of pigmented bacteria, yeasts and moulds.

Canned foods are often subjected to spoilage by bacteria such as *Clostridium* and *Bacillus* which survive as spores during the canning process. Bulging of cans is usually the result of fermentation of sugars with the release of carbon dioxide. The food in swollen cans is often sour owing to the conversion of sugar to organic acids. Putrefaction of canned meats and fish results in the release of hydrogen sulphide which is undetectable until the can is opened when, in addition to the foul smell, the inside of the can is usually seen to be blackened because of the production of sulphides of tin.

(b) Preservation

Any product can be sterilized if suitably treated by heat or germicide and kept under sterile conditions. Unfortunately, for most foods, such drastic treatments are of little use as a means of preservation as irreversible changes in flavour, texture and colour make the food inedible. The type of treatment selected for preserving food has to take into consideration its existing microflora, the potential microflora and its physical and chemical constituents. For these reasons the methods used are numerous and include traditional as well as recent techniques:

(*i*) *Low temperature.* This not only discourages microbial growth but retards autolytic reactions in the food. Root vegetables and fruits keep well during winter because of the prevailing low temperature. CHILLING or refrigeration at temperatures around freezing point is useful for preserving most perishable foods over several weeks. FREEZING at temperatures well below 0°C is now increasingly used since the advent of home freezers. Whilst preservation may be successful over a long time many foods lose texture under such severe conditions. Frozen food should be eaten immediately thawing has occurred as microbes which were dormant during the period of freezing multiply rapidly as the temperature rises. Refreezing of thawed food is not recommended because next time the product is thawed the microbial population, already given one chance to multiply, will grow again giving a very high level of contamination. A dangerous source of food poisoning is poultry which has not been properly thawed before it was cooked. Hens, ducks and turkeys are frequently contaminated with *Salmonella* bacteria which survive in poorly cooked parts of the carcass and multiply when the meat is cooled.

(*ii*) *High temperature.* This is a method used to kill micro-organisms in food, thereby sterilizing it. Two basic treatments are BOILING and PASTEURIZATION. Boiling is frequently employed in the canning of foods but it inevitably brings about considerable change in texture, flavour and colour of the product. Pasteurization is widely used for sterilizing milk and fruit juices. This process destroys about 99% of bacteria without altering the flavour or texture of the product. Two forms of pasteurization

101

are used today—the holding method and the high temperature method. The former method is similar to that used by Pasteur to eliminate undesirable bacteria from vats of grape juice. It consists of maintaining the product at a temperature of 63°C for half an hour before cooling. In the high temperature method the product is held at 72°C for fifteen seconds and then cooled.

(*iii*) *Dehydration*. This is the name given to the removal of water. It is a very useful method of food preservation because microbial growth is impossible without water. A second benefit of dehydration is the considerable loss in weight of the product which greatly reduces the cost of its transportation. The housewife has long been familiar with many dried foods such as peas, rice, and dried fruit. Such foods were often dehydrated by drying them in the sun. Today much of the drying of foods is factory based, and although many dried products are still prepared by placing them in hot air, newer techniques have been developed for some commodities. Vegetables, meat, poultry and fish are now often freeze-dried. They are frozen and placed in a chamber from which the air is evacuated when the moisture passes directly from the ice state to water vapour which is drawn off. Liquid foods such as skimmed milk and eggs are dried by spraying them into a jet of hot air. Powdered foods such as these tend to absorb atmospheric moisture so they have to be packed in air-tight containers.

As with freezing, dehydration does not kill microbes present on or in the food. Bacteria often multiply in dried food when water is added so it is wise to use the reconstituted food straight away. Dried egg, for example, is sometimes a source of food poisoning by the bacterium *Salmonella*.

(*iv*) *Preservatives*. These are chemicals added to food to discourage microbial growth. The use of preservatives is one of the oldest methods of food preservation. The curing or smoking of meat and fish and the use of spices which have mild germicidal properties, inhibits microbial activity. An additive with great microbial toxicity, but without harmful effects to man and causing no change in flavour or aroma to food, has yet to be found. Chemicals such as sodium benzoate and sulphur dioxide are widely used in low concentrations. Food laws however forbid the use of certain compounds which are known to be toxic to man.

(*v*) Foods with a high OSMOTIC POTENTIAL are unsuitable for microbial growth because any water present is bound to solute molecules and is therefore unavailable to micro-organisms. If a fungal or bacterial spore germinates on such foods, water is withdrawn from the germ tube by osmosis so the microbe quickly dies. The salting of meat and fish is a long practised method of food preservation. Jam and honey owe their keeping qualities to the high concentration of cane sugar they contain. Condensed milk is preserved partly by the increase in lactose achieved during the evaporation stage and also to the added sucrose.

(c) Food poisoning

A few toadstools such as the death cap agaric, *Amanita phalloides*, are so poisonous that a piece one cubic centimetre in size can kill a human being. Usually however food poisoning is caused by the presence of specific bacteria or their toxic products in or on ingested food. There are several types of bacterial food poisoning:

(*i*) *Botulinism.* This is a very dangerous form of food poisoning. Fortunately today it rarely occurs. It results from eating food containing an exotoxin produced by the bacterium *Clostridium botulinum.* This toxin is thought to be the most potent poison known; one ten thousandth of a milligram will kill a man. The toxin is a protein which has a powerful effect on the nervous system.

The bacterium normally lives in soil, but can also grow in food, without altering its appearance or flavour. Preserved foods, meat products and protein-rich vegetables are possible sources of infection. Several cases of botulinism have resulted from eating home canned foods where inadequate boiling failed to kill the bacterium. On cooling, the anaerobic conditions in the food stimulates spore germination.

(*ii*) *Staphylococcal.* This is the most common form of bacterial food poisoning and is caused by ingestion of an endotoxin produced during growth of *Staphylococcus aureus.* The victim has vomiting and diarrhoea, gastroenteritis, from which full recovery is made in about 48 hours. Resistance to the effects of the toxin is common, although, as all travellers will know, 'holiday tummy' often results from consuming food containing endotoxins chemically different from those one is used to at home. There is seldom any change in appearance or flavour of the infected food which will contain large numbers of the bacterium. Foods such as milk products, custards, creams, sauces and salted meats are likely sources of contamination.

(*iii*) *Salmonella.* This poisoning in its early stages resembles the less serious staphylococcal pattern of gastrointestinal infection. It occurs within four to twelve hours of eating contaminated foods such as meats, poultry, fish and their related products. Large numbers of bacteria must be present before poisoning occurs, and it is the presence of the bacterial cells rather than any toxin they produce which is responsible for the body reactions. Deaths from salmonella poisoning are less than 1% and recovery occurs within two to three days, although complications such as appendicitis or persistent fever may set in. One major problem lies in the fact that some patients may fully recover yet remain healthy carriers of the pathogen. If such persons are associated with the production or preparation of food, they serve as a continual focus of infection for future outbreaks.

8.5 SEWAGE DISPOSAL AND WATER PURIFICATION

In modern societies faecal material, urine, domestic and industrial liquid wastes are carried away through a system of sewers to sewage treatment plants. Sewage consists of 99·9% water and about 0·02% suspended solids, the remainder being soluble organic and inorganic compounds. The volume of sewage to be dealt with is enormous; in Britain, for example, 180 litres (40 gallons) daily for each person.

(a) Sewage treatment

The treatment of sewage involves the separation of solids from the water and the microbial breakdown of the solids to a harmless sludge which is released into streams and rivers or which is dried and used as a fertilizer (Fig. 8.5, Plate 8.4). Following screening of coarse materials such as paper, vegetable matter and grit, the sewage enters sedimentation tanks (Fig. 8.5A) where coarser solids settle out, forming a turbid liquid with a slightly unpleasant smell; already microbial activity has brought about chemical changes in the organic components. This process constitutes the PRIMARY TREATMENT.

Fig. 8.5 Outline of the activated sludge sewage process.

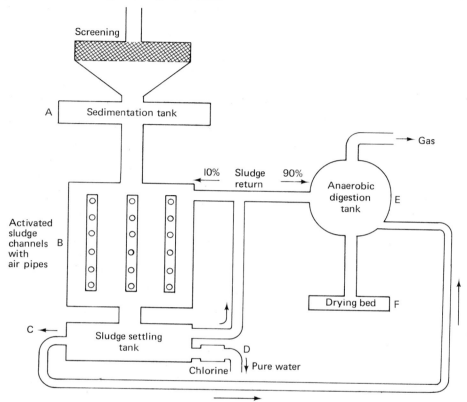

Plate 8.4 Activated sludge tanks (anaerobic digestion tanks in background)

The SECONDARY TREATMENT is entirely microbiological and involves aerobic bacteria working on the sewage which is circulated around large tanks (Fig. 8.5B) through which air is pumped. Stimulated by oxygen, the bacteria bring about rapid decomposition of the sewage during the few hours it stays in the tanks. The micro-organisms concerned include species of *Bacillus, Pseudomonas, Proteus* and the curious floc-forming bacterium *Zoogloea ramigera*. A healthy activated sludge will also contain protozoa such as the ciliate *Vorticella,* and the microbiologist can keep a check on the behaviour of the sludge by regular microscopical examination of the dominant populations. Since the majority of these micro-organisms are very sensitive to toxic chemicals, careful control over the discharge of harmful industrial effluent must be maintained. Synthetic detergents also present a problem as they are non-biodegradable. They persist and produce a foam which reduces aeration thereby inhibiting microbial activity.

From the activated tanks, the sludge enters settling tanks (Fig. 8.5C) where, following chlorination (Fig. 8.5D), relatively pure water can be removed to nearby water courses. Settled sludge, as a floc, is continually removed, about 10% returning to the aerated tanks to maintain inoculum size, the remainder going into anaerobic digestion tanks (Fig. 8.5E). Here, anaerobic bacteria, for example, *Methanobacterium,* carry out rapid digestion and form gases such as methane, which are burned to supply power to operate the plant:

$$4H_2 + CO_2 \rightarrow CH_4 + 2H_2O$$

Finally the digested sludge is sent to the drying beds (Fig. 8.5F) where it forms a cake, which may be used as a fertilizer with a high nitrogen value.

(b) Purification of water

Sterile water is seldom, if ever, found in nature. Water from highland streams will contain as few micro-organisms as one to ten per cubic centimetre. An important factor determining survival in water is the amount of available oxygen. Fast flowing rivers and streams are well aerated by their turbulence. With standing bodies of water such as pools and lakes, however, lack of oxygen below the surface seriously hinders microbial activity.

Hot water effluent from factories running into a river can produce a local temperature rise which reduces the solubility of oxygen in the water. Such thermal pollution, at the same time, increases the metabolic activity of all aquatic organisms thereby increasing the demand for oxygen.

Another major problem facing any micro-organism is the quantity of available nutrients. Only those species adapted to grow at low nutrient concentrations will survive in relatively pure water. These include bacteria of the genera *Pseudomonas* and *Achromobacter.* Enrichment of water with organic effluent such as raw sewage causes a dramatic increase in microbial numbers. The additional materials are said to have an oxygen demand as they stimulate growth of decomposer micro-organisms which rapidly use up oxygen in the water. This is expressed as the biochemical oxygen demand (B.O.D.). The greater the organic load added to the water, the higher becomes

105

the B.O.D. It is estimated that the oxygen contained in 2,000 gallons of clean water is required for oxidation of one day's waste products from a human being.

If the pollutant is of faecal matter then there is a chance that pathogenic bacteria will be present in the water. Several human diseases such as cholera and typhoid fever are transmitted by faecal contamination of drinking water. Under crowded living conditions with poor sanitary arrangements, outbreaks of either disease are likely to occur. The pathogenic bacteria of cholera, *Vibrio*, and of typhoid fever, *Salmonella*, are often present in faecal-contaminated water in numbers too low to enable them to be readily identified. However, the common and relatively harmless intestinal bacterium *Escherichia coli* will be present in very large numbers. If this can be recognized and confirmed by laboratory tests such as the presumptive coliform test (Chapter 11), then faecal pollution of the water is proved.

Allowing water to stand in a reservoir for two to three weeks will bring about sedimentation of particles and a lowering in the number of micro-organisms, as the nutrient level declines. This is a natural, self-purification process. Filtration of the water through beds of sand or fine gravel successfully removes the bulk of contaminants. Bacteria, fungi and protozoa are associated with the decay of any organic matter present and grow as a slime layer on the filter beds where similar reactions occur to those involved in the purification of sewage effluent. The addition of 0·02—2 p.p.m. of chlorine to the water supply will effectively kill water-borne pathogens without making it unsuitable for drinking purposes.

Apart from the hazards to health in drinking contaminated water, the presence of certain micro-organisms can also seriously affect industry by introducing production problems and causing corrosion damage to chemical plant and machinery.

8.6 BACTERIAL AND VIRAL DISEASES OF MAN

Micro-organisms from all major groups may be responsible for diseases of higher animals including man, but most diseases in man are the result of bacterial and viral infections (Table 8.3).

Table 8.3 Some bacterial and viral diseases of man

Bacterial	Anthrax, brucellosis, diphtheria, plague, tetanus, tuberculosis, typhoid fever, whooping cough, venereal diseases
Viral	Chickenpox, common cold, influenza, measles, mumps, poliomyelitis, rabies, rubella (German measles), smallpox

The human body is continuously exposed to contamination by micro-organisms. However, very few of these are able to establish themselves successfully either on or in the body due mainly to the inhibitory effect of natural secretions such as lysozyme in tears, and also because they have to compete with COMMENSAL micro-organisms associated with the body. This commensal human flora is made up of bacterial species which are always present in particular sites in a healthy body:

(*i*) On the SKIN. *Staphylococcus albus* and *S. aureus*, which persist even after vigorous washing, probably play an important part in inhibiting potential pathogens from becoming established on the body surfaces. However, the staphylococci may give rise to minor skin infections such as boils and pimples.

(*ii*) In the RESPIRATORY TRACT. The moist mucus membranes of the nose and mouth and respiratory tract harbour large numbers of bacteria especially staphylococci, but some potential pathogens such as pneumococci may be permanent residents.

(*iii*) In the GUT. A huge bacterial population lives in the alimentary canal with its permanent presence of food. Except for the stomach, where acid gastric juices prevent bacterial growth, all areas, especially the small intestine and rectum, have a typical flora dominated by coliforms such as *Escherichia coli*. Approximately one third of the dry weight of human faeces consists of bacteria.

INFECTIOUS diseases are caused by micro-organisms. If transmission readily occurs large numbers of cases develop in a short time, causing an epidemic. Some diseases such as the common cold and influenza are described as endemic; they occur constantly rather than sporadically.

The study of a specific bacterial disease, typhoid fever, and a viral disease, poliomyelitis, will show how the microbiologist is involved in the identification of the causative organism and in the development of control measures.

(a) Typhoid fever

(*i*) The PATHOGEN. Bacteria closely related to the intestinal coliform group are responsible for the dangerous diseases called typhoid and paratyphoid fever. They are *Salmonella typhi* and *S. paratyphi*. So closely related are these intestinal species to other salmonellae, which may be responsible for diseases such as food poisoning, that identification by routine staining and physiological tests is almost impossible. Serological tests are necessary. Serum prepared from an animal inoculated with a known strain of *Salmonella*, so as to include specific antibodies, is combined with *Salmonella* isolated from a suspected typhoid patient. On the basis of the resulting antibody—antigen response it is possible to identify the type or strain of *Salmonella* under investigation (Fig. 8.6).

(*ii*) The DISEASE and its TRANSMISSION. Typhoid fever is characterized by high fever, slow pulse rate, much enlargement of lymphoid tissues, pink coloured spots on the abdomen and considerable irritation of the intestinal tract leading to diarrhoea and ulceration. Initially the bacteria may be isolated from the blood, but after ten to fourteen days the urine and faeces contain the pathogen in large numbers. Death occurs at this stage in about 15% of untreated cases.

Paratyphoid fever is a much less virulent form of the infection but both typhoid and paratyphoid have similar methods of transmission. Since the organisms are enteric, entry to the body is by eating food containing the bacteria or by drinking water contaminated by sewage containing large numbers of *Salmonella*. Both forms of the disease are widely distributed throughout the world, especially where sewage removal and treatment is not practised in a hygienic way; they also occur in modern towns and cities following natural disasters such as earthquakes. The Aberdeen corned beef typhoid epidemic of 1964 was shown to originate from cooling the heated cans of meat in unchlorinated river water at a South American process factory. Bacteria entered some leaky tins, developed and con-

Fig. 8.6 The preparation of active serum and identification of serotypes

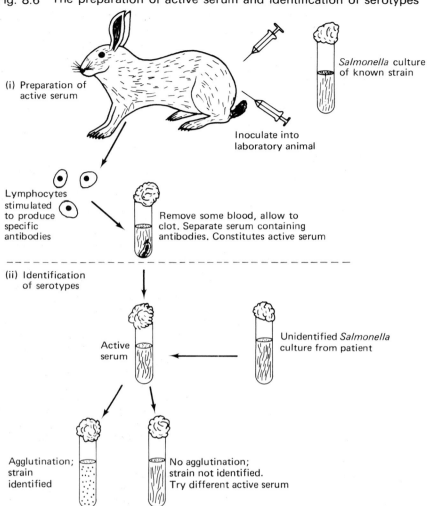

(i) Preparation of active serum

Salmonella culture of known strain

Inoculate into laboratory animal

Lymphocytes stimulated to produce specific antibodies

Remove some blood, allow to clot. Separate serum containing antibodies. Constitutes active serum

(ii) Identification of serotypes

Active serum

Unidentified *Salmonella* culture from patient

Agglutination; strain identified

No agglutination; strain not identified. Try different active serum

taminated the food, which finally ended its journey at an Aberdeen supermarket. Over 400 people consequently suffered from typhoid fever. Foods such as shellfish are likely to be sources of infection. These animals live in estuaries where sewage effluent is common. Since they feed by extracting bacteria and plankton from the water, they are likely to concentrate Salmonella within their tissues. Carriers, healthy individuals who carry *S. typhi* in their intestines, are likely to spread the disease especially if they work in the food industry. (*iii*) CONTROL of the disease. Since typhoid fever is associated with contaminated food and water, obvious preventative measures are the sanitary control of food production and the protection of water supplies by the proper disposal of human sewage.

Any infected individual should be immediately isolated and his faecal and urinary discharges disposed of with care following disinfectant treatment. Under the Public Health Regulations of 1953, carriers are prohibited from working in any occupation where food and drink is prepared or handled. Antibiotics such as chloramphenicol or tetracyclines can be used effectively to reduce the bacterial activity in the body and help recovery. It is now possible to vaccinate successfully against typhoid fever, such treatment giving immunity for periods of up to five years.

(b) Poliomyelitis (infantile paralysis)

(*i*) The PATHOGEN is a virus; the virion is a naked polyhedron (Chapter 5). It is one of the most stable of human viruses in that in pure culture it is highly resistant to disinfectants, antibiotics and other forms of harsh treatments. It is an enteric or intestinal virus, widespread in populations where a tolerance to its effects is part of the natural immune response reaction of the body. However, in a small proportion of infected persons, a form of the virus may occur which produces the visibly recognized symptoms of poliomyelitis, often leaving the victim with severe permanent disabilities.

(*ii*) The DISEASE and its TRANSMISSION. Poliomyelitis is a disease of the central nervous system which begins with symptoms such as vomiting and constipation followed by muscular spasms as the virions invade the nerve tissue. Temporary paralysis of the limbs may be produced at this stage but recovery may then slowly occur. However, in the event of the death of nerve cells a permanent paralysis of muscular tissue may result. Whilst all persons are at risk, the virus appears to induce damage in children more readily than adults, hence the name infantile paralysis. An attack of the disease generally gives life-long immunity. Since polio is essentially enteric, the viruses are excreted in faeces. Consumption of contaminated water and food is a likely means of contracting the disease. Bathing in faeces-contaminated river water can result in infection with poliomyelitis virus.

(*iii*) CONTROL of the disease. The most important method of control is that of active immunization which has reduced poliomyelitis to a relatively rare disease. Immunity against polio is induced by two vaccines, both of which evoke the body to produce antibodies against the virus and which remain at high levels in the blood serum for some years. In the Salk vaccine, introduced in 1953, formalin killed poliovirus is given by injection during the first year of life with periodic booster injections during early childhood. With the Sabin vaccine, a live but attenuated vaccine is used which is administered orally on a lump of sugar, and is believed to provide life-long immunity to the disease. Since immunity develops sooner and lasts longer with the Sabin treatment and the vaccine is cheaper to prepare and easier to administer, it is now preferred to the Salk vaccination.

8.7 FUNGAL AND VIRAL DISEASES OF PLANTS

A recent survey carried out in the United States of America estimated that plant diseases on average reduced annual crop yields there by 15–20%. A large proportion of this loss was due to viruses and pathogenic fungi (Table 8.4) even though control measures were widely practised.

Table 8.4 Some fungal and viral diseases of plants

Fungal	Apple scab, damping off of seedlings, downy mildews, ergot, Dutch elm disease, late blight, peach leaf curl, powdery mildews, vascular wilts, wheat rust
Viral	Beet yellows, cucumber necrosis, maize stunt disease, oat yellow leaf, pea yellow dwarf, potato leaf roll, tobacco mosaic, wheat streak mosaic

A knowledge of these parasites provides the basis for control measures and the role played by the plant pathologist must assume increasing importance if the growing world population is to be adequately fed.

(a) Late blight of potato

(*i*) The PATHOGEN is *Phytophthora infestans*, a phycomycete fungus belonging to the family *Pythiaceae*. It was responsible for the notorious famine among Irish peasants between 1845 and 1847 which reduced the population of Ireland, by starvation or emigration, by over two and a half million people. This provided an immense impetus to the study of plant diseases which has been sustained to the present day. Although the disease occurs wherever potatoes are grown it is now of secondary importance to viral diseases. Some varieties of potato, for example King Edward, are more susceptible to potato blight than are others such as Majestic.

(*ii*) The DISEASE and its TRANSMISSION. Late blight begins as small watery spots at the tips or margins of leaflets (Fig. 8.7). These lesions then become black in colour and shrivelled as cell death occurs. The infected areas spread to entire leaves along the petioles. Because much of the photosynthetic tissue is affected, translocation to the tubers is greatly reduced and the growth of the tubers is limited. Tubers become infected in the soil by spores washed down from blighted foliage and at harvesting when diseased haulms are present. Infected tubers possess small sunken watery spots at first which gradually turn brown and may be secondarily infected by saprophytic fungi and bacteria which rot the tuber.

The important features of the life cycle of the pathogen are as follows:

The fungus survives the winter in tubers infected the previous year. Although these tubers produce only a small proportion of infected shoots they act as focal points for the rapid spread of the disease. Sporangia are produced in large numbers and flicked off by hygroscopic twisting movements of the sporangiophores. They are carried by air currents to healthy leaves. Here they either germinate directly, that is behave as a conidium when the air temperature is above 18°C or, below 18°C give rise to biflagellate zoospores. Entry of germtubes into the host occurs either via stomata or directly into epidermal cells. Sporulation is very prolific when the humidity of the atmosphere is high. The dense leaf canopy in rows of potato plants encourages the development of a humid microclimate following a period of rain.

The sexual part of the life cycle occurs rarely in infected leaves or tubers and is unusual because the female organ, the oogonium, grows through the male organ, the antheridium, before a fertilization tube develops. An antheridial nucleus

110

Fig. 8.7 Life-cycle of *Phytophthora infestans*

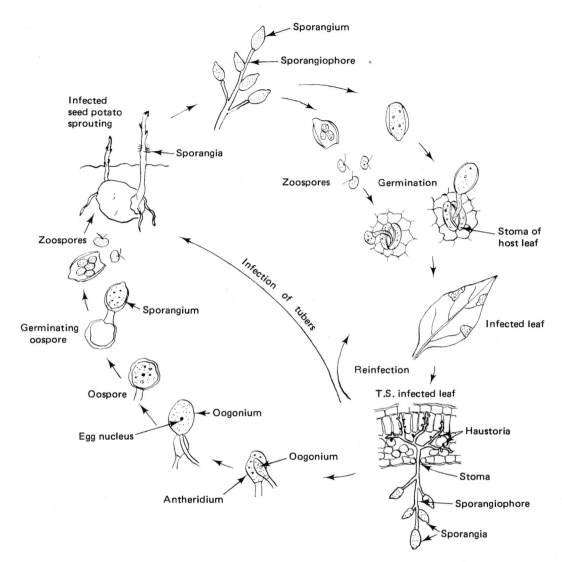

passes through the tube and fertilizes the egg nucleus. The egg now secretes a thick wall to become an oospore in which form the fungus can survive for long periods in the soil though it usually overwinters as mycelium in infected tubers.

(*iii*) CONTROL of the disease. The knowledge that blight is dependent on damp weather for its rapid spread and that infection in a crop begins with diseased seed potatoes can be put to advantage in controlling the disease. Blight is more severe in cool, wet weather because each sporangium then forms a number of zoospores instead of germinating directly as in warmer conditions. With the help of the Meteorological Office it is possible to predict the times of impending infections with some accuracy, following field observations on the first appearance of the disease. In Britain, ten- to fourteen-day warnings are now given in the main potato-growing regions. Protection can be given to the crop by spraying with a suitable fungicide to kill the germinating spores. Bordeaux mixture, a suspension of copper sulphate and calcium oxide, was once used extensively but now organic fungicides such as maneb, zineb and fentin hydroxide are preferred.

If certified, disease-free seed potatoes are used there is little chance of the disease establishing itself in a crop early in the year. Seed selected from previous diseased crops is more likely to contain the pathogen. Infected tubers should be destroyed or used as feedingstuff for animals.

Ungerminated sporangia die three to four weeks after their release. Tubers which are well covered with soil as they grow are less likely to become infected. The practice of ridging up soil at the base of the haulms helps in this respect.

Tubers may become infected from diseased haulms during lifting of the crop. By destroying the aerial shoots with sprays of a 10–15% solution of sulphuric acid, this form of infection is minimized. Because this operation is fraught with obvious dangers to the farmer, searches have been made for less corrosive defoliants but none of those tried so far have been very effective.

The ideal preventative measure would be to have varieties of potato which are immune to the pathogen. The wild potato, *Solanum demissum*, which is valueless as a crop plant, is known to be resistant to *P. infestans. S. demissum* has been crossed with the cultivated potato, *S. tuberosum* to produce resistant hybrids. Unfortunately, the fungus exists in at least 64 different physiological strains and no variety of potato yet bred is immune to all of them.

(b) Beet yellows virus

In this country beet yellows is a serious disease of sugar-beet, spinach and mangolds. The virus also infects common weeds such as groundsel, shepherd's purse and chickweed.

(*i*) The PATHOGEN. Thin sections of diseased leaf tissue display the virus dispersed throughout the cytoplasm or as compact inclusion bodies (Plate 8.5) which may fill the cell. The bodies consist of aggregations of virions. Each virion is a hollow filament measuring up to 1–2 μm long but only 10 mμ in diameter. The protein capsomeres are arranged helically. Virions have also been demonstrated in the pores of sieve plates, and in the protoplasmic threads, plasmodesmata, which connect sievetubes with adjacent parenchyma. This indicates the route by which the infection spreads in a diseased plant.

(*ii*) The DISEASE and its TRANSMISSION. The symptoms of the disease in sugar-beet first appear on the outer, older leaves which become yellow, brittle and thickened. Yellowing (chlorosis) always begins at the tip of the leaf and gradually spreads downwards between the veins. Death of tissues usually follows the chlorotic stage. Heavily infected leaves eventually become dry and crackle when crushed; the original name for the disease was crackly yellows. Internal physiological changes also occur in the phloem; the sugar content decreases and starch accumulates, which can lead to a 50% loss in the marketable value of the crop.

112

Plate 8.5 Thin section of a leaf cell containing virions, x5,000

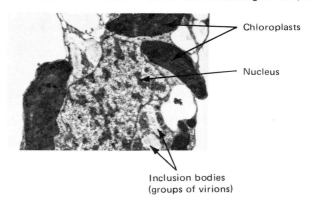

Chloroplasts

Nucleus

Inclusion bodies
(groups of virions)

Fig. 8.8 Transmission of beet yellows virus (modified after Dixon, 1973)

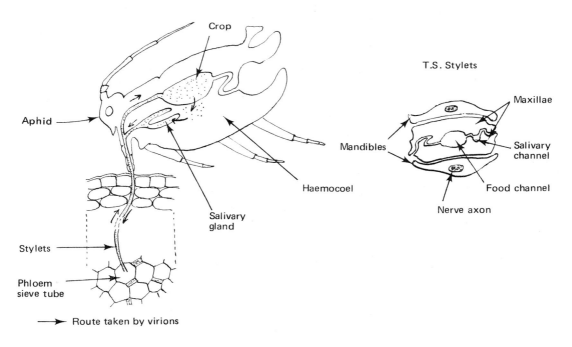

Crop

T.S. Stylets

Aphid

Maxillae

Mandibles

Salivary
channel

Haemocoel

Food channel

Salivary
gland

Nerve axon

Stylets

Phloem
sieve tube

⟶ Route taken by virions

The chief vectors of the disease are two species of aphid, the peach-potato aphid, *Myzus persicae*, the more important carrier, and the black-bean aphid, *Aphis fabae*. These feed by inserting their mouthparts, stylets, into plant cells and sucking up cell sap. The stylets comprise an outer pair of mandibles with sharp needle-like cutting points and an inner pair of maxillae. When not in use these are enclosed in a blunt *proboscis* but when the stylets are inserted into plant tissue the proboscis segments telescope into one another (Fig. 8.8).

113

While probing in the plant tissue a proteinaceous stylet sheath is secreted by the salivary glands to give flexibility to the mouth parts. Eventually a sieve tube is reached and cell sap, which is usually under considerable pressure in these cells, is forced up through the food canal. If the aphid is feeding on an infected plant there is a danger that some of the phloem sap will contain viral particles. Beet yellows virus is said to be CIRCULATIVE because following its passage to the aphid's stomach it enters the haemocoel and finds its way to the salivary glands where it replicates. The pathogen may then be transmitted to an uninfected plant. The ability to infect remains for long periods, often throughout the aphid's life. Moulting does not cause a loss of infectivity as it does in stylet-borne viruses where the infectious agent does not enter the body further than the mouth parts. Clearly there is a parallel between the transmission of beet yellows virus by aphids and that of the malarial parasite by mosquitos.

Peach-potato aphids move quickly from plant to plant thus spreading the infection quickly, but the black-bean aphid tends to be rather sedentary and is therefore less important in transmitting the disease (Fig. 8.9). The present day policy of growing large acreages of crops in monoculture encourages the rapid spread of any pathogen.

Fig. 8.9   Correlation between numbers of aphids and incidence of sugar beet yellows (after Watson, 1967).

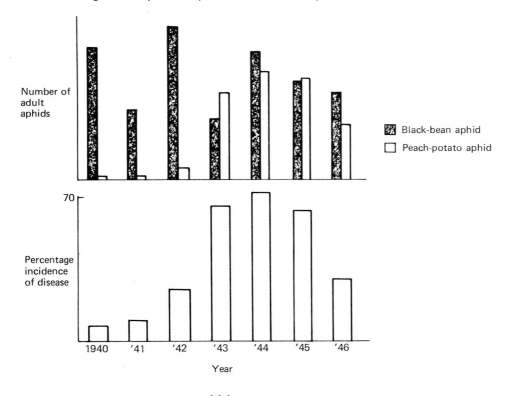

(*iii*) CONTROL of the disease. Sugar-beet is grown from seed which are obtained from stecklings, plants raised in summer to produce seed the following year. Both the root crop and stecklings are prone to infection but the disease is not transmitted via the seed. The spread of infection is entirely due to the activity of aphids. Adult aphids do not survive cold spells during winter and their eggs, which overwinter, do not contain the virus. Nymphs which hatch in spring must therefore become infected afresh. The vector multiplies rapidly by parthenogenesis and produces swarms of flying insects by June. With this knowledge it is possible to adopt a rational approach to eradication of the disease.

Aphicides can be used to protect both the root-crop and steckling beds from attack by the insects. Seeds used for raising stecklings can be treated with a systemic insecticide which gives the seedlings protection for about a month. Spraying of young steckling plants at two-week intervals until the end of October will keep them aphid-free over winter. They should receive a further treatment by the beginning of June when infestations with swarms of flying aphids begin. The root-crop should be sprayed as soon as there are signs of infestation with aphids, usually in June, but may be earlier after a mild winter. Sugar factory fieldmen observe the development of aphid populations and warnings are sent to sugar-beet growers when the need arises.

Sugar-beet and mangolds stored overwinter for cattle fodder are known to be potent sources of the virus. They should be used before April when the aphids begin their rapid rate of multiplication. It is impracticable to destroy all the alternative wild hosts of the vectors but it is advisable not to grow sugar-beet near to other aphid food sources such as brassicas and potatoes.

There is evidence that certain varieties of beet have an inherited resistance to aphids. If they can be developed for commercial use there is hope that the incidence of the disease will be greatly reduced. Virus-resistant strains of sugar-beet are also available. Their use has halved crop losses in East Anglia in recent years.

9 SAFETY IN THE MICROBIOLOGY LABORATORY

9.1 GENERAL PRECAUTIONS

You will have learned sufficient from the previous chapters to appreciate the fact that micro-organisms are constantly around us, on our body and clothes, in the air we breathe and on the laboratory bench and implements we handle. This means, of course, that to carry out microbiological experiments successfully, basic aseptic precautions must be maintained at all times. Far more important is the safety aspect. Although the microbes selected for the practical exercises in Chapter 11 are believed to be harmless to man, one can never be sure that pathogenic strains will not develop or pathogenic contaminants appear. It is **essential** therefore to treat **any** micro-organism as a potential pathogen and employ correct aseptic technique in its handling. This will ensure not only your safety but also that of your colleagues. The major source of infection in laboratories are AEROSOLS, tiny water droplets laden with bacteria released in the air as a result of careless working. These remain suspended in the atmosphere for long periods and may be inhaled. BREAKAGES and SPILLAGES of cultures may result in skin and eye infections whilst INGESTION of micro-organisms is very likely if cultures are pipetted by mouth.

The following simple laboratory rules should be observed to ensure safety in the laboratory:

1. Always wear a laboratory coat when working in the laboratory. This will not only prevent your clothes from being contaminated with the cultures you handle but will also protect your clothes from any stain splashes which may occur.
2. Ensure that any minor cuts or abrasions on exposed parts of your body are covered with a surgical dressing before participating in any experiment with microbes.
3. Make it a practice to wipe down the top of your bench with disinfectant at the beginning and end of each laboratory period. Any used cotton-wool or absorbent paper must be carefully discarded for burning.
4. Hand-to-mouth operations such as eating, smoking and the licking of labels should be forbidden. Moisten sticky labels with a drop of tap water and not with your tongue, or mark your specimen containers with a felt pen or wax pencil.
5. Never pipette **any** culture by mouth. Always use a teat pipette for transferring small amounts of liquid cultures. Do not forcefully squeeze the liquid from the pipette as you will create an aerosol.
6. If a culture is spilled report immediately to your supervisor. Should any accident occur, no matter how trivial, report it also to your supervisor.
7. When transferring micro-organisms ensure that the inoculating instrument is properly sterilized **before** and **after** the operation. Heat the wire to red heat on **each occasion.** Do not

flame loops which are heavily charged with inoculum as splattering with attendant aerosol formation will occur.

8. The opening of any container holding a culture will create an aerosol. Only open containers when asked to and do not sniff the cultures unless you are told that it is safe to do so.

9. Do not, under any circumstances, take any cultures out of the laboratory.

10. All used cultures, petri dishes or other glassware should be placed on a tray ready for autoclaving at the end of an experiment. Microscope slides used to prepare live mounts should be placed in disinfectant solution immediately after use.

11. Wash and dry your hands thoroughly before leaving the laboratory. Hot water, soap and throw-away paper towels should always be available.

9.2 DISPOSAL OF UNWANTED CULTURES AND HYGIENIC PRECAUTIONS

Used cultures should be quickly disposed of as they are a potential hazard to anyone working in the laboratory and often become a source of contamination to other cultures.

1. Plastic petri dishes containing agar cultures can be autoclaved in a suitable container such as a small metal bucket or in a sealed autoclavable bag (Sterilin Ltd.) and then disposed of via the refuse bin.

2. Cultures grown in glassware can be autoclaved and the vessels then washed for further use. After autoclaving pour the warm liquid medium down a sink and flush with hot water.

3. Immerse used microscope slides in a beaker containing 2% Chloros.

4. Place used pipettes immediately in jars filled with 2% Chloros ready for autoclaving.

5. Spillages of cultures in the laboratory are best dealt with by swabbing the contaminated area with a 10% solution of Chloros (Lysol is not recommended as it is not sporicidal, and besides it is a strong irritant).

6. Treat spillages of cultures on clothing with 1% Cetavlon (ICI).

9.3 SAFETY REFERENCES

Collins, C. H., Hartley, E. G., and Pilsworth, R. (1974). Prevention of laboratory acquired infection. *Public Health Laboratory Monograph No. 6.* HMSO.

Fry, P. J. (1975). Micro-organisms. *Schools Council Educational Use of Living Organisms Project.* English University Press.

Holt, G. (1974). Practical tips for the safe handling of micro-organisms in schools. *School Science Review 56* **(195):** 248–52.

Wray, J. (1974). Recommended practice for schools relating to the use of living organisms. *Schools Council Publication.* English University Press.

The use of micro-organisms in schools (1977). *DES Pamphlet No. 61.* HMSO London.

10 THE CULTIVATION AND EXAMINATION OF MICRO-ORGANISMS

10.1 EQUIPMENT

The basic equipment needed for growing microbes includes:

10.1.1 Containers

in which sterile media are kept ready for use and in which cultures are grown (Fig. 10.1). Those which are used most frequently are:

(*i*) *Petri dishes.* Sterile plastic dishes are now readily available and these are indispensable in any microbiology laboratory. They are used to grow microbes on solid agar media and then disposed of. Glass petri dishes which can be re-used after sterilization are also available.

(*ii*) *Bacteriological tubes* are used to hold 10–15 cm³ of solid or liquid media. They are just long test tubes with a loose metal cap or a cotton wool bung which can be removed with ease when transferring an inoculum. Agar media may be prepared as a deep or as a slope; the particular uses of each will be explained later.

Fig. 10.1 Containers used for the growth of micro-organisms.

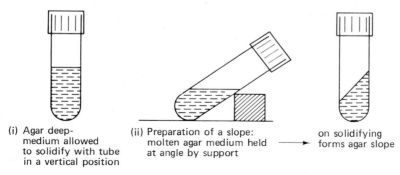

(i) Agar deep-
   medium allowed
   to solidify with tube
   in a vertical position

(ii) Preparation of a slope:
   molten agar medium held
   at angle by support

on solidifying
forms agar slope

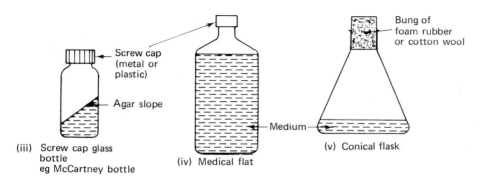

Screw cap
(metal or
plastic)

Agar slope

(iii) Screw cap glass
   bottle
   eg McCartney bottle

(iv) Medical flat

Medium

Bung of
foam rubber
or cotton wool

(v) Conical flask

(*iii*) *Screw cap glass bottles* of 20–30 cm³ capacity such as McCartney bottles. These can be used instead of bacteriological tubes but beginners may experience difficulty in removing the cap quickly enough during inoculation procedures.

(*iv*) *Medical flats* are flat glass medicine bottles; 100 cm³ and 250 cm³ sizes are most useful. The main function of these is to store large quantities of prepared sterile media. Such volumes of agar media take about an hour to melt. However, they do save time when a large number of agar plates have to be poured. Do not try to cool agar quickly in these bottles by running cold water over them as they are very prone to cracking.

(*v*) *Conical flasks*, preferably 100 and 250 cm³, are required for growing fungi in liquid culture media.

10.1.2 Inoculating instruments

(Fig. 10.2). Because the wires have to be sterilized at red heat in a bunsen flame it is wise to have instruments with metal handles. Glass handles crack very easily when inoculating wires and loops are heated in a bunsen flame.

Fig. 10.2 Inoculating instruments used in microbiology.

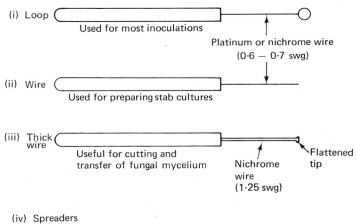

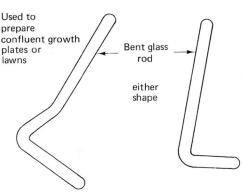

10.1.3 Pipettes

Sterile pipettes are essential for many microbiological exercises. Small disposable sterile pipettes can be purchased but there is no problem in sterilizing glass pipettes. Graduated 1, 5, 10 and 20 cm³ sizes are required most frequently. Pasteur pipettes are also often used (Fig. 10.3). All pipettes should have a small piece of cotton wool pushed into the mouth as a dust filter.

Fig. 10.3 A Pasteur pipette.

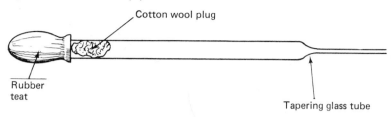

Cotton wool plug

Rubber teat

Tapering glass tube

10.1.4 An autoclave or a pressure cooker

This is needed to sterilize media and equipment such as glassware that will withstand high temperatures. If microbiological work is done regularly an autoclave is much preferred to a pressure cooker as large quantities of materials can then be sterilized at one time. Other methods such as steaming and boiling are time consuming and do not always work.

Metal containers in which pipettes can be housed during autoclaving are available. When sterilizing media in screw cap containers enough room must be left inside the bottle for the medium to expand and the cap partially unscrewed so that air can escape.

10.1.5 Incubators

are needed to grow cultures at known temperatures. Ideally several should be available to provide a range of temperatures for some investigations.

Plate 10.1 Effect of temperature on the growth of *Fusarium*, x $\frac{2}{9}$

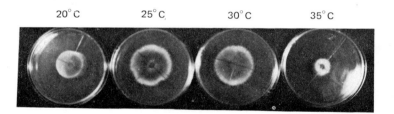

20°C   25°C   30°C   35°C

10.1.6 Inoculating cabinets

are highly desirable when any transfer of microbes is being carried out. They protect the operator from the microbes he is handling and also minimize contamination of the cultures from air-borne particles. Expensive cabinets in which air is drawn or blown over the cultures are available but a relatively cheap yet effective cabinet can be purchased from Philip Harris Biological Ltd.

10.1.7 Microscopes

are essential for observing micro-organisms. At least one or two should be fitted with an oil immersion objective for demonstra-

120

tions. Less expensive microscopical accessories such as micrometers and haemacytometers should be available in larger quantities. While it is useful to have phase-contrast facilities, these are not essential for microbiological work.

10.2 PREPARATION OF MEDIA

Many media capable of supporting the growth of microorganisms are now available in a ready-prepared, dehydrated state (Oxoid Ltd., and Difco). It is often convenient to dissolve the ingredients in water contained in a large beaker or flask, and then dispense the resulting medium into appropriate containers ready for autoclaving. Medical flats should be only partially filled, leaving an air gap of 20–30 cm^3, while 10–15 cm^3 aliquots are dispensed into McCartney bottles or bacteriological tubes. Liquid media used for certain fungal growth experiments are usually autoclaved in conical flasks, 20–30 cm^3 for a 100 cm^3 flask, 50–75 cm^3 for a 250 cm^3 flask, stoppered with cotton wool bungs.

10.2.1 Types of media

The liquid media used for bacteriological work are generally known as BROTHS. Solid media are prepared by adding agar to the basic broth ingredients prior to autoclaving. On cooling, the agar, which is of no nutritional value, sets to give a jelly-like consistency to the medium.

NUTRIENT MEDIA are widely used in bacteriological work. They can be made as follows:

(*i*) NUTRIENT BROTH
1. Dissolve 10 g meat extract or LabLemco, 10 g of peptone and 5 g of sodium chloride in 1 l of tap water. Heat gently if necessary.
2. Adjust the medium to pH 7·4 using MNaOH or MHCl and dispense into suitable containers such as McCartney bottles.
3. Autoclave at 103·4 kN/m^2 (15 psi) for 15 minutes after ensuring that bottle tops are only lightly screwed on.
4. On cooling screw down the bottle caps firmly.

(*ii*) NUTRIENT AGAR
Nutrient agar is prepared like nutrient broth but 20 g of plain agar is dissolved in the mixture before autoclaving. After sterilization when the medium is still molten, agar slopes can be prepared if required (Section 10.1).

Three of the most common media used for fungi are potato dextrose agar, malt agar and Czapek–Dox medium.

(*i*) POTATO DEXTROSE AGAR—PDA
1. Wash a number of sound potato tubers, cut into slices and boil 200 g in tap water until soft.
2. Crush the boiled tuber tissue in the water and when cool strain through clean muslin.
3. Make up the filtrate to 500 cm^3 with distilled water and add 20 g glucose.
4. Melt 20 g of plain agar in another 500 cm^3 of water and mix with the potato–glucose solution.
5. Dispense into suitable containers and autoclave at 103·4 kN/m^2 for 15 minutes.

Because of the lack of uniformity between different batches of potatoes, PDA is unsuitable for comparative physiological work. Nevertheless it is frequently used for morphological studies.

121

(ii) MALT AGAR

1. Dissolve 15 g of agar in 1 l of water and add 20 g of malt extract.
2. Dispense into appropriate vessels and autoclave as for PDA.

This medium is again more often used for morphological studies. Unmodified malt agar will encourage the development of ascocarps of *Penicillium wortmanni* while *Aspergillus repens* will produce sexual fruiting bodies if 20 g sucrose is added per 100 cm^3 of the medium.

Saccharomyces cerevisiae can be induced to form asci by first growing the species on nutrient agar containing 2 g glucose/ 100 cm^3 until white glistening colonies appear. These are then used to inoculate a sporulation medium of pH 6·5–7 containing 3–5 g anhydrous sodium acetate and 20 g agar in 1 l of water on which the yeast is incubated for 24–48 hours at 25°C.

(iii) CZAPEK–DOX MEDIUM

1. Dissolve 0·01 g of hydrated zinc sulphate in 500 cm^3 of distilled water then add 0·5 g magnesium sulphate (hydrated), 2 g sodium nitrate, 0·5 g potassium chloride, 0·01 g ferrous sulphate (hydrated), 1 g of potassium phosphate (K_2HPO_4 or KH_2PO_4), 30 g sucrose and 0·005 g copper sulphate (hydrated).
2. Melt 15 g of plain agar in another 500 cm^3 distilled water and mix with the salt–sugar solution.
3. Autoclave as for PDA.

If the agar is omitted a liquid medium is prepared. Because the precise composition of the medium is known it is especially useful for physiological studies.

The preparation of other media will be described at appropriate places in the practical exercises in Chapter 11.

Fig. 10.4 Pouring an agar plate.

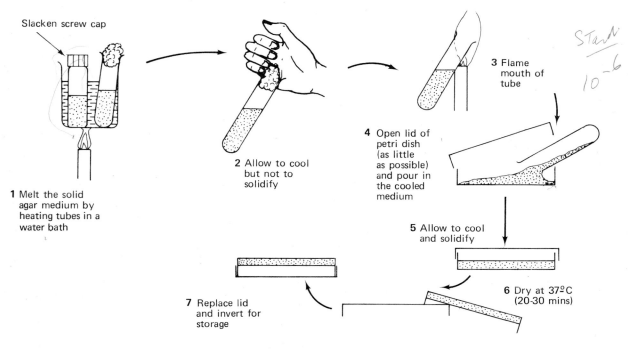

Slacken screw cap

1 Melt the solid agar medium by heating tubes in a water bath

2 Allow to cool but not to solidify

3 Flame mouth of tube

4 Open lid of petri dish (as little as possible) and pour in the cooled medium

5 Allow to cool and solidify

6 Dry at 37°C (20-30 mins)

7 Replace lid and invert for storage

10.2.2 Pouring an agar plate

Plates of agar media already poured can be purchased (Oxoid Ltd.) but only a few media are available in this form. The procedure shown in Fig. 10.4 is adopted in the laboratory. Larger volumes of agar media can be stored in 250 cm³ medical flats and melted down when required. It will take about an hour to melt the medium in these but you can pour about 15–20 plates from each bottle.

10.3 INOCULATION OF MEDIA

In all experimental work with micro-organisms it is important to use young cultures otherwise typical results are not obtained. Always sterilize inoculating instruments before and immediately after use.

10.3.1 Bacteria

(*i*) STREAK PLATES. This technique is often used for separating mixed cultures into pure ones because the individual colonies

Fig. 10.5 (a) Preparing a streak plate.

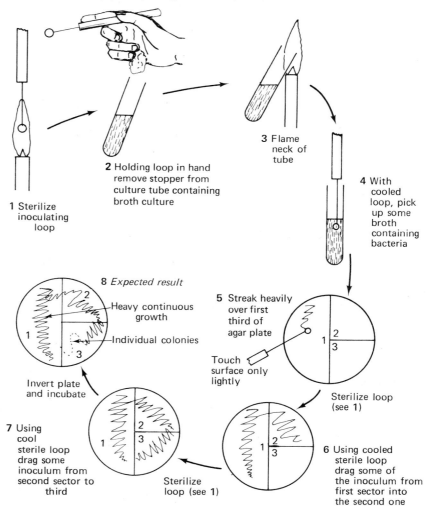

1 Sterilize inoculating loop

2 Holding loop in hand remove stopper from culture tube containing broth culture

3 Flame neck of tube

4 With cooled loop, pick up some broth containing bacteria

5 Streak heavily over first third of agar plate

Touch surface only lightly

Sterilize loop (see 1)

6 Using cooled sterile loop drag some of the inoculum from first sector into the second one

Sterilize loop (see 1)

7 Using cool sterile loop drag some inoculum from second sector to third

Invert plate and incubate

8 *Expected result*

Heavy continuous growth

Individual colonies

Fig. 10.5 (b) Streaking a plate with pure cultures.

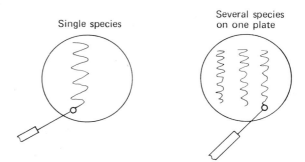

Single species

Several species
on one plate

which become separated can be used to inoculate fresh plates
or tubes. When pure cultures are used in physiological studies
an alternative method of streaking is used (Fig. 10.5).
(*ii*) POUR PLATES are prepared as shown in Fig. 10.6. Each colony
develops from a single bacterium present in the inoculum so
this method is suitable for counting the number of bacteria in a
sample of liquid.

Fig. 10.6   Preparing a pour plate.

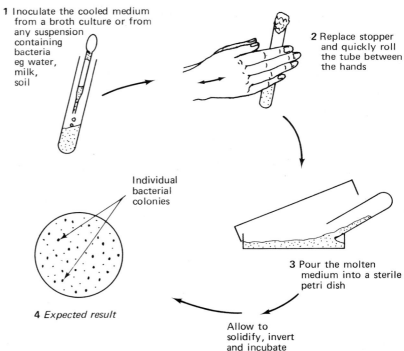

1 Inoculate the cooled medium
from a broth culture or from
any suspension
containing
bacteria
eg water,
milk,
soil

2 Replace stopper
and quickly roll
the tube between
the hands

Individual
bacterial
colonies

3 Pour the molten
medium into a sterile
petri dish

4 *Expected result*

Allow to
solidify, invert
and incubate

(*iii*) AGAR SLOPES are inoculated with aerobic bacteria as shown
in Fig. 10.7.
(*iv*) AGAR DEEPS are used to culture anaerobes or for separating
aerobes from anaerobes. The inoculation technique is shown in
Fig. 10.8.

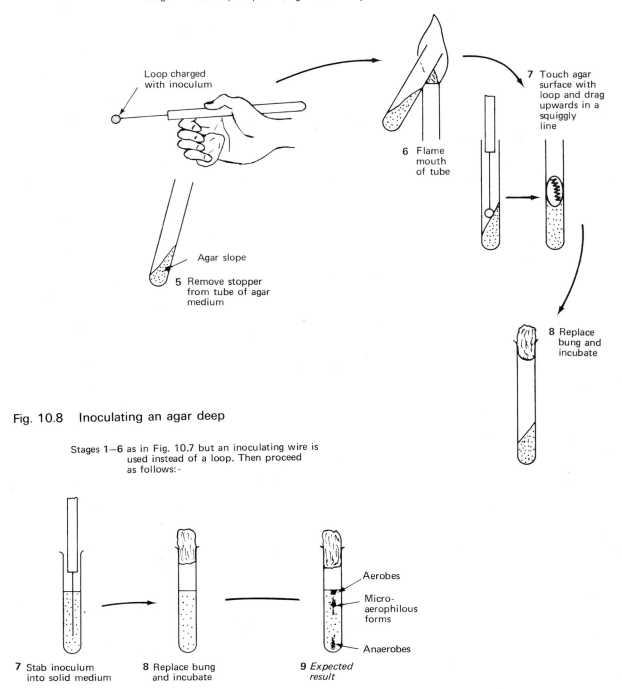

Fig. 10.7   Inoculating an agar slope.

Stages **1—4** as for petri plates (Fig. 10.5) Then proceed as follows:-

Loop charged
with inoculum

7 Touch agar
surface with
loop and drag
upwards in a
squiggly
line

6 Flame
mouth
of tube

Agar slope

5 Remove stopper
from tube of agar
medium

8 Replace
bung and
incubate

Fig. 10.8   Inoculating an agar deep

Stages **1—6** as in Fig. 10.7 but an inoculating wire is
used instead of a loop. Then proceed
as follows:-

Aerobes

Micro-
aerophilous
forms

Anaerobes

7 Stab inoculum
into solid medium

8 Replace bung
and incubate

9 *Expected
result*

125

(*v*) BROTHS are inoculated as for agar slopes. The charged loop is swirled in the broth in order to transfer the inoculum.

(*vi*) CONFLUENT PLATE or LAWN. For some exercises such as microbiological assays a dense sheet of bacterial growth over the surface of the medium is needed. This is achieved as shown in Fig. 10.9.

Fig. 10.9   Preparation of a confluent plate

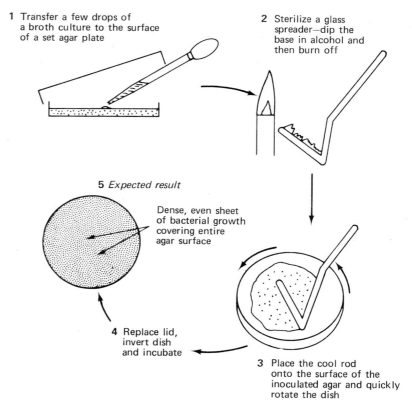

1 Transfer a few drops of a broth culture to the surface of a set agar plate

2 Sterilize a glass spreader—dip the base in alcohol and then burn off

5 *Expected result*

Dense, even sheet of bacterial growth covering entire agar surface

4 Replace lid, invert dish and incubate

3 Place the cool rod onto the surface of the inoculated agar and quickly rotate the dish

10.3.2 Fungi

Spores or hyphae can be used as sources of inocula, the choice depending on the nature of the investigation and on the species used. Appropriate methods will be given in the practical exercises in Chapter 11.

(*i*) MYCELIAL DISCS are used to inoculate an agar plate as shown in Fig. 10.10. Discs may also be transferred to liquid media.

(*ii*) SPORES. The inoculation of an agar plate with fungal spores is shown in Fig. 10.11. Liquid media can be inoculated by swirling the charged loop in the nutrient solution.

10.4 STOCK CULTURES

Although it is possible to purchase cultures of many microbes it is always useful to have a small collection available for immediate use. Details of the collection should be catalogued in a book with the culture number, source and date of acquisition alongside the name of the species. Stock cultures of many fungi and bacteria are best maintained on agar slopes which are labelled with the name of the species, catalogue number, the medium and date of last transfer. The bottles should be stored with their lids unscrewed half a turn or so. Most species can be kept alive for six to twelve months in a refrigerator on suitable media. The longevity of fungal and bacterial cultures on solid media can be increased to several years by pouring sterile liquid paraffin into the bottle until the culture is completely immersed. These bottles should then be stored in a cool, dry situation which is free from draughts. A cardboard box with a loose-fitting lid is ideal.

Fig. 10.10   Inoculation of an agar plate with a mycelial disc

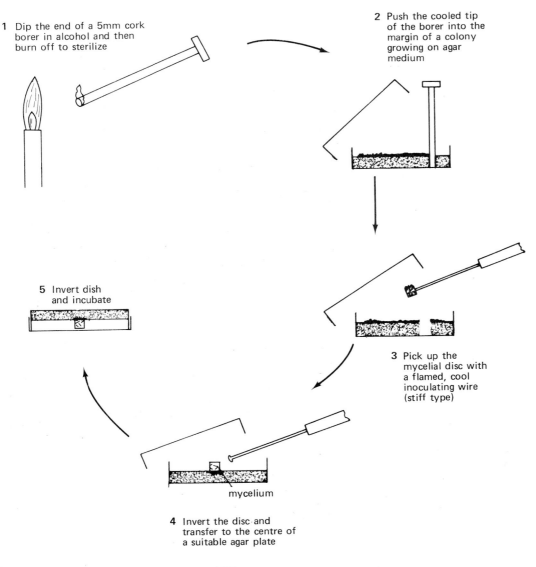

1 Dip the end of a 5mm cork borer in alcohol and then burn off to sterilize

2 Push the cooled tip of the borer into the margin of a colony growing on agar medium

3 Pick up the mycelial disc with a flamed, cool inoculating wire (stiff type)

4 Invert the disc and transfer to the centre of a suitable agar plate

mycelium

5 Invert dish and incubate

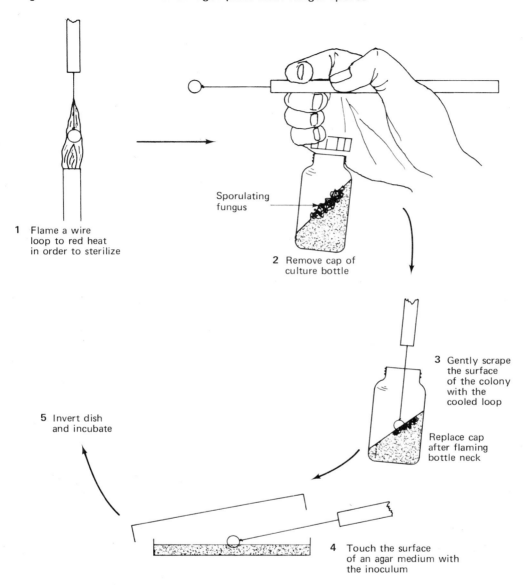

Fig. 10.11   Inoculation of an agar plate with fungal spores

1 Flame a wire
loop to red heat
in order to sterilize

Sporulating
fungus

2 Remove cap of
culture bottle

3 Gently scrape
the surface
of the colony
with the
cooled loop

Replace cap
after flaming
bottle neck

5 Invert dish
and incubate

4 Touch the surface
of an agar medium with
the inoculum

10.5 STAINING METHODS

10.5.1 Bacteria

Bacteria can be observed in an unstained state. This is usually
unsatisfactory for colourless species whose refractive index is
little different from that of water, thus making it difficult even
to find them. The use of stains accentuates the difference in
refractive index of the bacterium and its surroundings. Bacteria
can be stained leaving the background colourless (POSITIVE
STAINING), or unstained bacteria can be suspended in a dark
medium (NEGATIVE STAINING). Gram in 1884 devised a positive

staining technique which is still widely used in preliminary tests for the identification of unknown species (Chapter 2).

Fig. 10.12   Preparation of a bacterial smear

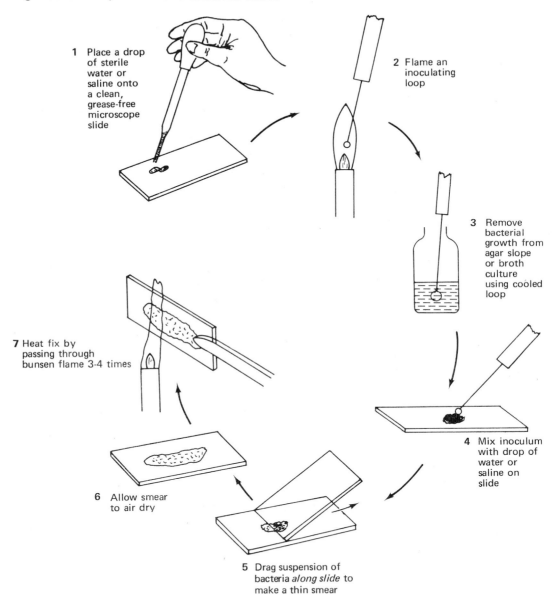

1 Place a drop of sterile water or saline onto a clean, grease-free microscope slide

2 Flame an inoculating loop

3 Remove bacterial growth from agar slope or broth culture using cooled loop

4 Mix inoculum with drop of water or saline on slide

5 Drag suspension of bacteria *along slide* to make a thin smear

6 Allow smear to air dry

7 Heat fix by passing through bunsen flame 3-4 times

(*i*) GRAM'S STAINING METHOD. A thin smear of the bacteria is first prepared (Fig. 10.12). The heat-fixed smear is then stained as follows:

1. Flood the cool smear with 1% aqueous crystal violet stain for 1–2 minutes.

2. Wash with tap water then flood with Gram's iodine for one minute.

3. Decolorize the smear with acetone or absolute alcohol. To do this run the solvent quickly over the smear until no more stain is released.

4. Wash with tap water and counterstain with dilute carbol fuchsin or 1% aqueous safranin for 2–3 minutes.

5. Wash with tap water and dry with blotting or filter paper. Gram positive bacteria retain the violet stain when decolorization is attempted so they appear blue or purple in the smear. Those which are Gram negative lose the violet stain then take up the second dye to colour red or pink in colour.

SUGGESTED ORGANISMS: *Bacillus subtilis* and *Aerobacter aerogenes*.

PREPARATION OF GRAM'S IODINE. Dissolve 1 g iodine crystals and 2 g potassium iodide in 300 cm³ of distilled water.

(*ii*) STAINING OF CAPSULES. The chemical composition of capsules varies from one bacterial species to another so there is no positive staining technique which is applicable to all capsulate species. The following negative staining procedure is generally satisfactory:

1. Place a loopful of Indian ink on to a clean microscope slide.

2. Mix into this a loopful of a broth culture or a scraping from a culture grown on an agar medium.

3. Place a coverslip on top of the suspension and press down gently with a piece of blotting paper.

Examine under a microscope. The capsule appears as a clear halo between the cell wall and the dark background.

SUGGESTED ORGANISMS: *Azotobacter agilis, Aerobacter aerogenes*.

(*iii*) STAINING OF SPORES—CONKLIN'S METHOD

1. Prepare a heat fixed smear of a young culture of a sporing bacterium.

Fig. 10.13   Heating of smear for spore stain

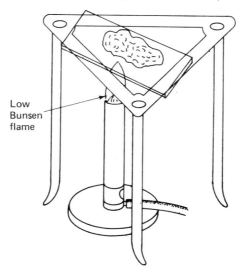

Low Bunsen flame

2. Flood with 5% aqueous malachite green and heat gently for 5–10 minutes. Add fresh stain when necessary to prevent the smear from drying out (Fig. 10.13).

3. Allow to cool and wash with tap water.

4. Flood with 0·25% aqueous safranin for 30–60 seconds.

5. Wash with tap water and blot dry.

Spores will be stained green and the vegetative parts of bacteria red.

SUGGESTED ORGANISMS: *Bacillus subtilis, B. stearothermophilus.*

10.5.2 Fungi

Fig. 10.14 Making a semi-permanent slide preparation of a fungus

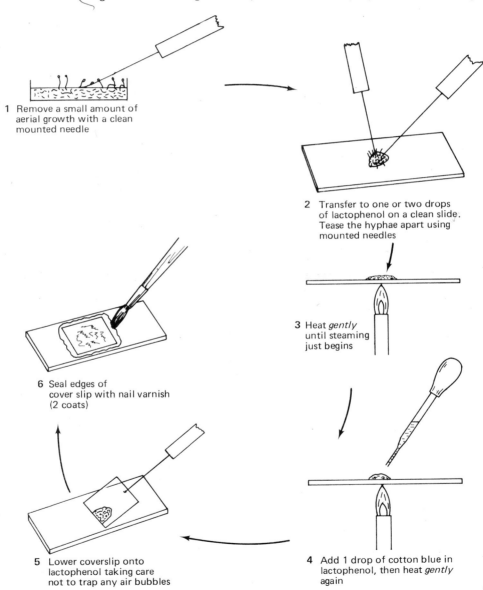

1 Remove a small amount of aerial growth with a clean mounted needle

2 Transfer to one or two drops of lactophenol on a clean slide. Tease the hyphae apart using mounted needles

3 Heat *gently* until steaming just begins

4 Add 1 drop of cotton blue in lactophenol, then heat *gently* again

5 Lower coverslip onto lactophenol taking care not to trap any air bubbles

6 Seal edges of cover slip with nail varnish (2 coats)

Diseased tissue or plate cultures of some of the smaller fungi can be observed using the low power objective of a microscope or a stereomicroscope without further manipulation. However, some species grow so densely that essential morphological features cannot be discerned, or further magnification may be needed. In such cases it is necessary to make slide preparations (Fig. 10.14). Those species which do not develop coarse aerial growth are best prepared using small agar blocks carefully cut from the colony. The block may have to be warmed and squashed when the coverslip is applied. Thin sections of diseased tissue can be stained using this method.

10.6 MICROSCOPICAL TECHNIQUES

10.6.1 Measuring the size of micro-organisms

This is achieved with a pair of micrometers. One of these, the eyepiece micrometer, is placed on a ledge inside the eyepiece of the microscope (Fig. 10.15). When in position the scale on the micrometer is superimposed on the image of the specimen when the observer looks down the microscope. The size, diameter, length, or width of the specimen is measured in eyepiece units, EPU. However the actual magnitude of each EPU is unknown and must now be determined using a stage micrometer. This is a glass slide on which a scale of known dimensions is etched. A frequently used type has divisions of 0.1 mm (100 μm).

Fig. 10.15 Eyepiece and stage micrometers

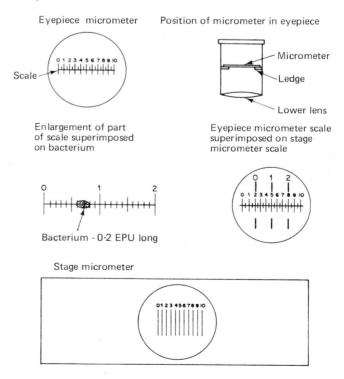

Eyepiece micrometer

Scale

Position of micrometer in eyepiece

Micrometer

Ledge

Lower lens

Enlargement of part of scale superimposed on bacterium

Bacterium - 0·2 EPU long

Eyepiece micrometer scale superimposed on stage micrometer scale

Stage micrometer

The stage micrometer is placed on to the microscope stage and the scale is found with the same eyepiece and objective used to measure the specimen originally, in our example high power $\times$ 10. Now the eyepiece scale is superimposed on the scale of the stage micrometer.

$$\text{We can see that 3 EPU (2-5)}$$
$$\text{equals 1 slide division or 100 } \mu\text{m}$$
$$\therefore 1 \text{ EPU} = \frac{100}{3} = 33\cdot3 \, \mu\text{m}$$
$$\text{The bacillus measured 0·2 EPU long}$$
$$0\cdot2 \text{ EPU} = 33\cdot3 \times 0\cdot2 = 6\cdot6 \, \mu\text{m}$$

Calibrate the eyepiece micrometer using each of the dry objectives and the oil immersion lens. Keep a note of the calculations to save time on future occasions.

10.6.2 Measuring the number of cells in a liquid sample

A special counting chamber known as a haemacytometer which was originally designed to count blood cells can be used for this purpose (Fig. 10.16). The chamber has a central platform on which is etched a 1 mm square grid divided into 400 small squares. The side of each small square is thus $\frac{1}{20}$ mm and the area of each is $\frac{1}{400}$ mm.

Fig. 10.16 A haemacytometer

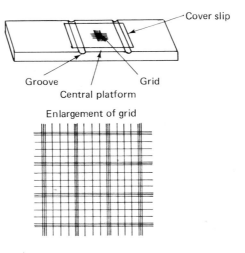

The edges of the coverslip are gently pressed on to the side platforms until a rainbow effect, Newton's rings, appears. The distance between the central platform and the undersurface of the coverslip is now exactly 0·1 mm so that the volume of suspension above each small square is $\frac{1}{400} \times 0\cdot1 = \frac{1}{4000}$ mm^3. A drop of the microbial suspension is placed on the edge of the coverslip so that it runs on to the grid. The number of cells lying above at least 80 small squares is now counted. Those lying over the top and right sides of the squares are included but those lying on the bottom and left hand sides are ignored.

If, for example, we counted 200 cells present these would be suspended in $\frac{80}{4000}$ mm^3 of liquid.

Thus the number present in $1\ mm^3 = 200 \times \dfrac{4000}{80} = 10{,}000$.

Samples containing very large numbers of cells may be diluted but the dilution factor must then be accounted for in the final calculation.

10.6.3 Observing bacteria for motility

The hanging drop method is generally used (Fig. 10.17).

Find the edge of the drop under the low power objective. This is not difficult if the light is not too intense. Now turn the high power objective into position. Make sure that you can distinguish between true motility and Brownian movement. The latter is displayed by all small particles in aqueous suspension because they are continually bombarded with water molecules. This causes a random but local movement of the particles. Motile bacteria on the other hand move considerable distances quite quickly and will be seen to dash across the field of view.
SUGGESTED ORGANISMS: *Spirillum serpens, Micrococcus agilis.*

Fig. 10.17 Preparation of a hanging-drop microslide

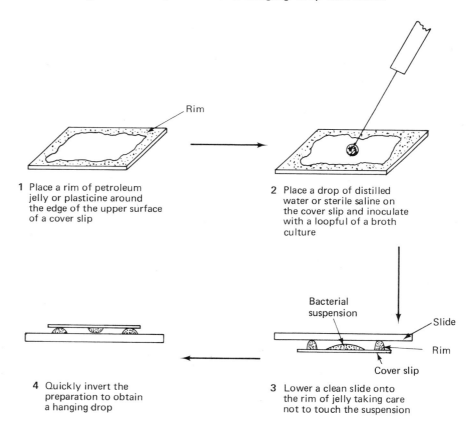

1 Place a rim of petroleum jelly or plasticine around the edge of the upper surface of a cover slip

2 Place a drop of distilled water or sterile saline on the cover slip and inoculate with a loopful of a broth culture

4 Quickly invert the preparation to obtain a hanging drop

3 Lower a clean slide onto the rim of jelly taking care not to touch the suspension

11   PRACTICAL EXERCISES

11.1 INVESTIGATION OF THE RANGE OF MICRO-ORGANISMS IN NATURAL HABITATS

Pure cultures of many micro-organisms can be purchased but are expensive. As an introduction to the variety of micro-organisms it is an easy matter to prepare mixed cultures of unknown organisms using infusions. There are innumerable ways of making these but hay infusions often prove successful:

1. Place 3 or 4 pieces of clean, dry hay or lawn cuttings into a clean petri dish.

2. Add enough pond or stream water to cover the hay. Replace the lid and keep at room temperature for 1–2 weeks away from direct sunlight. Top up with more pond water if necessary. Other sources of material are liquid from the rumen of a slaughtered animal and stagnant water.

3. Using a clean Pasteur pipette transfer a drop of liquid from near the hay to a microscope slide. Add a coverslip and examine under the low power objective. Use the high power objective later if required.

Questions

1. Can you identify representatives from some of the major groups of micro-organisms in your preparation? Make annotated drawings to display the major characteristics of the structure and behaviour of some of them.

2. Measure the size of an organism from each major group.

136

11.2 DEMONSTRATION OF THE WIDESPREAD OCCURRENCE OF MICRO-ORGANISMS

Even the cleanest laboratory is contaminated with microbes which are suspended in the air, present in dust on the benches, floor and equipment, in the water supply and on or in the users of the laboratory. The distribution of some of these micro-organisms can be demonstrated as follows:

1. Pour 10–15 cm³ of cool, molten nutrient agar into each of six petri plates.
2. Allow to solidify and then use the plates as follows:
(a) Remove the lid from the first plate and leave the agar in direct contact with the air for 20–30 minutes.
(b) Touch the agar surface in the second dish with your fingers.
(c) Wash your hands thoroughly, dry them with a clean towel and repeat (b) with a third plate of agar.
(d) Cough or sneeze on to the surface of the agar in a fourth dish.
(e) Moisten a piece of sterile cotton wool with some sterile water and use it to swab a small area of the bench surface. Now touch the surface of the agar in a fifth dish with the swab.
(f) Place 1 cm³ of tap water on to the agar surface of a sixth dish. Use a sterile spreader to make a thin film of the water on the surface of the medium.
3. Replace the lid of the dish immediately each operation is completed. Invert the dishes and incubate at 25°C for several days.
4. Examine the plates and describe the appearance of the colonies. Note the size, shape, colour and texture of each growth. Do not open plates while examining the colonies.

Questions

1. What methods can be used to control microbial contaminants in the laboratory?
2. How could the investigation be improved to isolate other contaminating micro-organisms?

11.3 NUTRITION

11.3.1 Production of extra-cellular enzymes

A. AMYLASE SECRETION. Some microbes produce amylase enzymes which hydrolyse the glycosidic linkages of starch molecules. By flooding an agar medium containing starch, with iodine in potassium iodide solution, after microbial growth it is possible to see whether digestion of the polysaccharide has occurred.

Fungal amylases

1. Pour plates of starch agar.
2. When the agar has set inoculate each plate centrally with an inverted mycelial disc of one of the test fungi.
3. Incubate at 25°C and when the colony is well established pour a 0·2% aqueous solution of iodine in potassium iodide over the medium.
4. After 2–3 minutes wash away the excess iodine and note the colour of the medium.
SUGGESTED FUNGI: *Mucor*, *Chaetomium*, *Sordaria* and *Fusarium*.

Preparation of starch agar

(i) Dissolve 3 g sodium nitrate, 0·5 g potassium chloride, 0·5 g hydrated magnesium sulphate, 1 g disodium hydrogen phosphate, 0·01 hydrated ferrous sulphate and 15 g of agar in 900 cm³ of distilled water.
(ii) Heat 2 g of starch in 100 cm³ of water to form a suspension. Allow to cool and then mix with the agar mineral solution.

(*iii*) Autoclave at 103·4 kN/m² for 15 minutes.

Bacterial amylases

1. Using the bacterial cultures provided prepare streak plates (Fig. 10.5) on nutrient agar containing 1 g of soluble starch per 100 cm³ of medium.
2. Incubate for 2–3 days at 30°C then flood the plates with dilute iodine solution.
3. After 2–3 minutes pour the excess iodine solution into disinfectant and note the colour of the medium.
SUGGESTED ORGANISMS: *Micrococcus luteus, Bacillus subtilis, Escherichia coli* and *Streptococcus faecalis.*

Questions

1. What indication is there that some of the starch has disappeared in the vicinity of the growth?
2. Write an empirical equation to summarize the hydrolysis of a molecule of starch.
3. What is the likely fate of the products of hydrolysed starch?

B. CELLULASE SECRETION BY FUNGI
1. Pour plates of cellulose agar after shaking the medium to disperse the cellulose particles.
2. When the agar has set inoculate each plate centrally with an inverted disc of a test fungus.
3. Incubate at 25°C and when the colonies are well established hold the plate near to a strong light and see whether there is any change in opacity of the medium where growth has occurred. Alternatively, carefully scrape away the mycelium and compare the appearance of cellulose particles where the fungus has grown with those in the uncolonized agar.
SUGGESTED TEST FUNGI: *Sordaria, Chaetomium, Fusarium* and *Myrothecium.*

Questions

1. What evidence is there for the disappearance of cellulose from the medium?
2. What is the ecological significance of cellulose decomposition?
3. How important are bacteria in the decay of cellulose in nature?
PREPARATION OF CELLULOSE AGAR:
(*i*) Dissolve 0·5 g ammonium sulphate, 0·5 g L-asparagine, 1 g potassium dihydrogen phosphate, 0·5 g potassium chloride, 0·2 g hydrated magnesium sulphate, 0·1 g calcium chloride and 0·5 g yeast extract in 1 l of distilled water.
(*ii*) Add 10 g of powdered cellulose (SIGMACELL TYPE 20) and 20 g agar.
(*iii*) Autoclave at 103·4 kN/m² for 15 minutes.

C. SECRETION OF PEPTIDASE ENZYMES BY BACTERIA. These enzymes hydrolyse peptide linkages within protein molecules. The clearing of milk in milk-agar plates is regarded as an indication of casein hydrolysis provided the plates are developed with acidic mercuric chloride solution. **Take care,** this is a poisonous reagent. It precipitates undigested protein, causing the medium to become opaque, leaving clear zones where the casein has been broken down.
1. Pour plates of milk agar (5 cm³ sterile milk added to 100 cm³ cool, sterile aqueous suspension of 2% agar).
2. When the agar has solidified streak the plates with the test

organisms (Fig. 10.5).

3. Incubate at 30°C for 7 days.

4. Flood the plates with acid mercuric chloride (15 g mercuric chloride, 20 ml concentrated hydrochloric acid, 100 cm³ distilled water).

5. Compare the opacity of the medium in the vicinity of growth with that of the remainder of the plate.

SUGGESTED ORGANISMS: *Escherichia coli, Bacillus subtilis, Lactobacillus lactis, Proteus vulgaris.*

Questions

1. Which of the ingredients of the medium provides the casein?

2. What are the likely products of casein hydrolysis?

3. What is the probable fate of these products?

4. Write an equation to illustrate the cleavage of a peptide linkage.

D. SECRETION OF LIPASE ENZYMES BY BACTERIA. The formation of clear zones adjacent to microbial growth on a medium containing tributyrin is an indication of lipase activity. This test is used in food bacteriology for detecting the presence of lipolytic bacteria in butter.

1. Pour plates of tributyrin agar (1 g tributyrin in 100 cm³ nutrient agar).

2. When the agar has solidified streak the plates with the test organisms (Fig. 10.5).

3. Incubate at 30°C for 3–4 days and examine for signs of clearing around the streaks of bacterial growth.

SUGGESTED ORGANISMS: *Escherichia coli, Proteus vulgaris, Bacillus cereus.*

Questions

1. What is tributyrin and what are the likely products of its hydrolysis?

2. What changes are likely to occur in butter or milk containing lipolytic bacteria?

3. In what way are lipid-digesting microbes of importance in the purification of water?

11.3.2 Demonstration of the production of vitamins by fungi (after Dade and Gunnell, 1969)

1. Inoculate a plate of Czapek–Dox agar with a mycelial disc of *Sordaria fimicola.*

2. Inoculate a second plate at one end with *S. fimicola* and at the other end with a mycelial disc of *Aspergillus rugulosus.*

3. Incubate at 25°C for 7–10 days or until there is evidence of the production of perithecia by *S. fimicola.*

4. Remove perithecia and squash in a drop of lactophenol. Observe under the microscope.

Questions

1. Does *S. fimicola* produce perithecia on both plates?

2. Where are the perithecia produced in the mixed culture?

3. Are the perithecia fertile on both media? What evidence can you provide to substantiate your answer?

4. *S. fimicola* requires the vitamin biotin to produce fertile perithecia. This is not present in the agar medium so where is it derived from when fertile fruiting bodies are produced?

11.3.3 Investigation of the mineral requirements of a fungus

1. Prepare a complete mineral liquid medium such as Czapek–Dox (C–D), medium using deionized water (section 10.2.1). Also prepare media lacking each of the essential elements (see below).

2. Dispense 25 cm³ aliquots of each medium into separate 100 cm³ conical flasks. Plug with cotton wool and autoclave at 103·4 kN/m² for 10 minutes.

3. Gently scrape the surface of a sporulating culture of a fungus with a moist sterile inoculating loop and transfer the spores to 10 cm³ of sterile water in a McCartney bottle. Shake thoroughly to disperse the spores.

4. Inoculate the cooled flasks of medium with 0·5 cm³ of spore suspension.

5. Incubate at 25°C for 7–14 days.

6. Note any differences in sporulation of the fungus in the various media then harvest the mycelia by filtration.

7. Dry to constant mass at 100°C and weigh the amount of growth produced in each medium.

SUGGESTED ORGANISMS: *Penicillium, Aspergillus.*

Questions

1. On which medium did the fungus grow best of all?

2. Was there any correlation between dry mass and the amount of sporulation?

3. What are the roles of nitrogen, sulphur and phosphorus in the metabolism of micro-organisms?

4. What are the functions of those elements which do not participate in the formation of structural components within microbial cells?

INGREDIENTS FOR DEFICIENT MEDIA:

(*i*) Minus sodium. Substitute potassium nitrate for the same weight of sodium nitrate in C–D.

(*ii*) Minus potassium. Substitute sodium chloride for the same weight of potassium chloride in C–D.

(*iii*) Minus magnesium. Substitute ferrous sulphate for the same weight of magnesium sulphate in C–D.

(*iv*) Minus iron. Omit ferrous sulphate.

(*v*) Minus phosphorus. Omit potassium phosphate.

(*vi*) Minus nitrogen. Omit sodium nitrate.

(*vii*) Minus sulphur. Substitute ferric chloride for the same weight of ferrous sulphate and magnesium chloride for the same weight of magnesium sulphate.

(*viii*) Minus trace elements. Omit zinc and copper sulphates.

11.4 RESPIRATION

11.4.1 Respiratory enzymes

A. DEHYDROGENASES. The activity of dehydrogenases can be demonstrated by the use of dyes such as methylene blue which act as artificial hydrogen acceptors and become colourless on being reduced. The methylene blue test is widely used in the dairy industry to assess the microbial activity and therefore the degree of contamination of milk and ice cream.

1. Shake up a sample of milk and transfer 10 cm³ to a sterile McCartney bottle using a sterile pipette.

2. Add 1 cm³ of standard methylene blue solution (1:20,000 aqueous solution), screw the cap on tight, invert once to mix and place in a water bath at 37°C.

3. Meanwhile prepare two control tubes as follows:

(*a*) 10 cm³ milk + 1 cm³ tap water

(*b*) 10 cm³ milk + 1 cm³ methylene solution

} boiled at 100°C for 3 minutes.

These serve as comparisons to show when the decolorization begins and when it is complete.

4. Examine the experimental tubes at 30-minute intervals. Tubes in which the sample is completely decolorized can be removed, others are inverted once to mix the sample and then replaced. Tuberculin tested milk should not be decolorized within $4\frac{1}{2}$ hours in summer or $5\frac{1}{2}$ hours in winter. Pasteurized milk should not be decolorized within 30 minutes.

Reduction of the dye may be speeded up if small samples of natural yoghurt are added to the milk before the test is started.

Questions

1. What is the role of dehydrogenase enzymes in living organisms?
2. What is the reason for the reduction of the methylene blue?

B. OXIDASES. In the method which follows tetra methyl-p-phenylenediamine is oxidized to a blue compound by bacterial oxidases.

1. Prepare streak plates on nutrient agar using the broth cultures provided.
2. Incubate at 30°C for 5 days.
3. Add a few drops of 1% tetra methyl-p-phenylenediamine in 0·1% ascorbic acid. If oxidase is produced the bacterial growth will turn blue within one minute.

This test is used to distinguish between members of the *Pseudomonadaceae* and *Enterobacteriaceae*.

SUGGESTED ORGANISMS: *Pseudomonas fluorescens, E. coli, Bacillus subtilis, Micrococcus luteus.*

Questions

1. What is the role of oxidases in living organisms?
2. Would you expect to find oxidases in anaerobic bacteria? Give reasons for your answer.

C. CATALASE. Many bacterial oxidases catalyse the formation of hydrogen peroxide instead of water when oxidizing reduced hydrogen carriers in aerobic conditions. As the peroxide is toxic it is quickly reduced to water and oxygen by catalase. Do not carry out this test on microscope slides as the effervescence creates aerosols which may be a source of contamination.

1. Place 1 cm³ of a 3% aqueous hydrogen peroxide (10 volume) into a clean screw cap bottle.
2. Using a sterile inoculating loop, remove a small amount of bacterial growth from an agar slope culture.
3. Mix the bacteria with the peroxide, replace the screw cap and watch for the production of bubbles. Use a fresh tube of peroxide for each species. Some cultures will produce considerable frothing, others may have to be observed closely for evidence of bubbling.

SUGGESTED ORGANISMS: *Escherichia coli, Streptococcus faecalis, Staphylococcus albus, Lactobacillus acidophilus.*

Questions

1. Why does the hydrogen peroxide froth when mixed with some species of bacteria?
2. Would you expect catalase positive bacteria to be aerobes or anaerobes? Give reasons for your choice.

141

11.4.2 Investigation of the oxygen requirements of bacteria

1. Stab-inoculate nutrient agar deeps (15–20 cm³ agar per tube) with the organisms provided, using a fresh deep for each species.
2. Incubate at 30°C for 7 days.
3. Observe and make a note of the location of growth of each species.

SUGGESTED ORGANISMS: *Escherichia coli, Aerobacter aerogenes, Lactobacillus acidophilus, Mycobacterium phlei.*

Questions

1. To what categories in terms of oxygen requirements do (*i*) those growing at the surface of the medium, (*ii*) those growing just below the surface and (*iii*) those growing at the bottom of the medium belong?
2. By what process is oxygen made available in the cultures to those bacteria needing the gas?
3. How could you modify this investigation to separate aerobic bacteria from a mixture of aerobes and anaerobes?

11.4.3 Fermentation

A. FERMENTATION OF SUGAR IN DOUGH BY BAKER'S YEAST. (after Bottle, 1967).
1. Mix 100 g of plain flour with 5 g of glucose.
2. Disperse 7 g of fresh baker's yeast, *Saccharomyces cerevisiae*, in 120 cm³ of water at 45°C and stand the mixture in a warm place for 10–15 minutes.
3. Stir the yeast suspension into the glucose/flour mixture to produce a smooth paste.
4. Pour the paste into a 1 l measuring cylinder without allowing the mixture to touch the sides of the cylinder. There is no need to transfer all of the mixture.
5. When the dough has settled note the volume it occupies. This is zero time.
6. Record the volume at five-minute intervals until the skin on the dough shows signs of rupture. Plot a graph of volume against time.

Questions

1. Why does the volume of the dough increase with time?
2. Write an empirical equation to summarize the process of fermentation by yeast.
3. What increase in volume would you expect if the 5 g of glucose was completely fermented? How does this compare with your results? Suggest reasons for any discrepancy.
4. What results would you expect if you performed the investigation at different temperatures?

EXTENSION WORK: The investigation can be carried out at different temperatures by standing the cylinders in water baths at 20, 25, 30 and 35°C. In addition the mass of yeast and of glucose may be varied in order to investigate the effect of enzyme and substrate concentrations on the rate of fermentation.

B. ALCOHOLIC FERMENTATION OF FRUIT JUICE. Wine yeast, *Saccharomyces cerevisiae var. ellipsoideus,* and fruit juice concentrate can be obtained from shops which stock home-brewing materials.
1. Place 50 cm³ of sterile reconstituted raisin, grape or prune juice in a flask and inoculate with 1 cm³ of a broth culture of wine yeast.
2. Incubate at 25°C for 7 days.

3. Note the smell of the product.
4. Place a drop of the sediment on to a microscope slide, add a drop of Gram's iodine, and examine for budding yeast cells.

Questions

1. Why is the yeast respiring anaerobically?
2. Write an equation to summarize the fermentation of glucose by yeast.
3. How is sugar prepared for fermentation in the brewing industry?
4. Make drawings to show successive stages in budding.
5. How does budding in a yeast differ from binary fission in a bacterium?

PREPARATION OF FRUIT JUICE: Reconstitute the concentrated fruit juice according to the instructions on the package.

PREPARATION OF INOCULUM: Stir 5 g of dried wine yeast in 100 cm³ of sterile 10% glucose solution and leave for 15–20 minutes.

11.5 GROWTH

11.5.1 Construction of a growth curve for a yeast

1. Prepare 15 McCartney bottles, each containing 15 cm³ of sterile Pasteur's solution.
2. Inoculate each tube with two drops of a suspension of growing yeast cells.
3. Incubate at 25°C and after an hour measure the cell density in three of the tubes using a haemacytometer (section 10.6.2). A colorimeter or better still a nephelometer may be used instead if available. Repeat with another three tubes after a further hour. Take further readings every two hours until all of the tubes have been used. See section 11.16 for an alternative approach.
4. Plot a graph of mean cell density, number of cells per cm³ of medium, against time.

Questions

1. What is being measured as an index of growth in this experiment?
2. What is meant by the term exponential growth? Over what period does it occur in your result?
3. Suggest reasons for the changes in growth rate during the course of the investigation.
4. Apart from the time factor, is the growth cycle of other organisms comparable with that of a yeast? If so, what factors are likely to limit the growth of human populations?

PREPARATION OF PASTEUR'S SOLUTION:
(*i*) Dissolve 15 g sucrose, 1 g ammonium tartrate, 0·2 g dipotassium hydrogen phosphate, 0·02 g calcium phosphate and 0·02 g magnesium sulphate in 100 cm³ of distilled water.
(*ii*) Autoclave at 103·4 kN/m² for 10 minutes.

PREPARATION OF THE YEAST INOCULUM: Disperse 5 g of fresh baker's yeast in 100 cm³ of sterile water at 45°C.

11.5.2 The effect of temperature on the growth rate of fungi

1. Inoculate plates of PDA centrally with a mycelial disc cut from the non-sporing margin of a fungal colony (Fig. 10.10).
2. Invert the dishes and incubate three replicates at temperatures of 5°C (refrigerator), 20, 25, 30, 35, and 40°C or whatever range of temperatures is available.
3. Measure two diameters at right angles to one another of each colony at 12, 24, or 48 hour intervals depending on the growth rate of the species.

4. Plot a graph of mean colony diameter against temperature after 7 days or other suitable time interval.

SUGGESTED ORGANISMS: Species which do not sporulate freely until growth is well established are preferred so that sterile mycelial discs can be used as an inoculum. This avoids the problem of scattering of spores during inoculation which results in a large number of small colonies growing in a dish. *Fusarium* and *Trichoderma* are useful subjects (Plate 10.1).

Questions

1. What is being measured as an index of growth in this investigation?
2. What conclusion can be drawn from your graph on the effect of temperature on growth of the fungi?
3. What is the temperature coefficient, Q_{10} between 20 and 30°C?
4. Do your graphs resemble that for the growth of a yeast or bacterial population?

PRECAUTION: At temperatures of 30°C and above there is a tendency for media to dehydrate rather quickly. Avoid this by placing a shallow container of water in the bottom of the incubator.

11.5.3 The effect of pH on the growth of microbes

1. Prepare tubes of buffered nutrient broth of pH 3·6, 5·0, 6·6 and 7·9.
2. Inoculate each tube with a loopful of broth culture of a bacterium. Inoculate another series with a loopful of broth culture of a yeast.
3. Incubate at 30°C for 5–7 days and then measure the cell density in each tube using a haemacytometer (section 10.6.2), colorimeter or nephelometer.
4. Plot a graph of cell density against pH.

SUGGESTED ORGANISMS: *Escherichia coli, Saccharomyces cerevisiae.*

Questions

1. What is meant by the term buffering?
2. Why are the broths buffered?
3. Suggest reasons why the pH of the environment affects the growth of micro-organisms.
4. Fungi are said to be acid-tolerant. In what habitats could this tolerance be exploited?

PREPARATION OF BUFFERED BROTHS: The medium is prepared by adding sterile buffer solutions to sterile nutrient broth in the quantities shown in Table 11.1.

Table 11.1

pH	cm³ broth	cm³ 0·1 M citric acid	cm³ 0·2 M dipotassium hydrogen phosphate
3·6	80	17	3
5·0	80	11	9
6·6	80	7	13
7·9	80	1	19

11.6 MICROBIOLOGY OF WATER

11.6.1 Counting the number of bacteria in a sample of water

The next exercise could be carried out at the same time as this one. Safety or teat pipettes should be used to transfer samples during the dilution procedure. Never examine open dishes of bacteria, because pathogens may be present.

1. Allow the cold water tap to run for a few minutes before collecting a sample of water in a sterile bottle.
2. Using a sterile pipette, transfer $10 \, cm^3$ of the collected water into a bottle containing $90 \, cm^3$ of sterile saline (0.9% NaCl solution) to produce a dilution of 1 in 10 (10^{-1}).
3. Shake well, and using a fresh sterile pipette transfer $10 \, cm^3$ of the 10^{-1} dilution to a second bottle containing $90 \, cm^3$ of sterile saline thus producing a dilution of 1 in 100 (10^{-2}). Do not throw away the 10^{-1} dilution.
4. Repeat the diluting procedure twice more to produce dilutions of 10^{-3} and 10^{-4}. Do not discard the 10^{-2} dilution.
5. Place $1 \, cm^3$ of each sample into separate petri dishes marked with the appropriate dilution. Carry out this step in duplicate.
6. Pour $15 \, cm^3$ of cool, molten nutrient agar into each dish and swirl the dish gently to mix the contents thoroughly.
7. Incubate one set of dilutions at 22°C and the other at 37°C for two or three days. Count the numbers of colonies on each plate. Do not open the dishes. Plates with less than 30 or more than 300 colonies should be ignored for counting purposes.
8. Multiply the number of colonies by the reciprocal of the dilution to find the number of bacteria in the original sample of water. For example, if there were 100 colonies on the plates made from the 10^{-2} dilution, the number of bacteria in the original sample = $100 \times 10^2 = 10,000/cm^3$.

Questions

1. How does each bacterial colony growing on the agar arise?
2. What is likely to be the origin of the bacteria growing on the plates incubated at the higher temperature? Give reasons for your answer.
3. Suggest ways in which the experiment could be improved to give a more accurate reflection of the numbers of bacteria in water.

11.6.2 Presumptive coliform test

Standard methods are used for determining the numbers of coliform bacteria in a water sample but some are time consuming. A simple method involves the use of Endo agar on which coliforms readily grow from serial dilutions of water. Endo agar contains lactose and reduced fuchsin among other things. Coliforms ferment the lactose to form deep red colonies with a golden sheen.

1. Prepare 10^{-2}, 10^{-3} and 10^{-4} dilutions of a sample of tap water as described in the previous investigation.
2. Using a teat pipette transfer $1 \, cm^3$ samples to separate petri dishes, marked with the appropriate dilutions. Carry out this step in duplicate.
3. Pour $15 \, cm^3$ of cool, molten Endo agar (Oxoid Ltd.) to each dish and swirl the dish gently to distribute the bacteria evenly in the medium.
4. Incubate one set of dilutions at 22°C and the other at 37°C.
5. Without opening the dishes, count the number of deep red colonies on each plate.
6. Multiply the number of coliform colonies by the reciprocal of the dilution (see 11.6.1) to find the number of presumed

coliforms in the water.

Questions

1. Why is the presence of coliforms in drinking water regarded as a potential danger?
2. What measures are taken to eliminate coliforms in the purification of water?
3. Why is it important not to open the dishes when you are examining them?

11.6.3 Isolation of aquatic moulds

Baiting techniques are widely used to isolate aquatic fungi from fresh water habitats.
1. Boil some ant or fly eggs for a few minutes or boil a few grass or hemp seeds until the seed coats split.
2. Transfer one or two pieces of bait to a large clean dish containing 50 cm³ of aerated stream, river or pond water.
3. Cover the dish and incubate at 25°C for 2—3 days.
4. Examine the baits under low power of the microscope for signs of fungal growth. Periodic aeration of the water helps to reduce bacterial contamination.

Questions

1. What substrates are colonized by aquatic fungi in their natural environment?
2. How do these organisms survive when food is not available?
3. What other factors are likely to limit the activity of aquatic fungi in their natural habitats?
EXTENSION WORK: Exercises 11.6.1, 11.6.2 and 11.6.3 may be used to compare the microbiology of water from a number of sources such as swimming pools, rivers, ponds, canals and the sea. Never examine open dishes of the bacteria as pathogenic organisms may be present.

11.7 MEDICAL MICROBIOLOGY

11.7.1 Investigation of the antibiotic sensitivity of bacteria

1. Prepare pour plates or lawns (section 10.3.1) of known bacteria such as *B. subtilis, E. coli, Staph. albus* and *Micrococcus luteus.*
2. Place a multisensitivity antibiotic disc (Oxoid Multodisk) on the surface of each plate and incubate without inverting the plates at 37°C for 2—3 days.
3. Measure the diameter of the inhibition zones around each of the discs for each organism.

Questions

1. Why is there a clear zone around each of the discs?
2. Which antibiotic is most effective against each of the species tested? Give reasons for your answer.

11.7.2 The effect of disinfectants on microbial growth

1. Prepare pour plates or lawns of known bacteria. When the agar has set mark the back of the dish into four sectors using a wax pencil.
2. Prepare a range of concentrations of common disinfectants such as Lysol, Chloros, Tego and Dettol in sterile water; 10%, 5%, 2·5% and 1·25% solutions are suitable.
3. Using sterile forceps, immerse a 5 mm sterile filter paper disc in each solution.
4. Shake off excess liquid and place discs containing different

146

concentrations of disinfectant on the centre of the agar in each quadrant of one of the plates. Mark the concentration and name of the disinfectant on the appropriate quadrant.

5. Incubate at 37°C for 1—2 days and then measure the diameter of the inhibition zone around each disc.

6. Construct a standard growth curve for each bacterium and each disinfectant.

Questions

1. Which of the disinfectants is most effective in inhibiting the growth of each species tested?

2. What factors are likely to govern the size of the inhibition zone for any one concentration of each disinfectant?

3. How would you use the method to assay an unknown concentration of disinfectant or antibiotic in solution?

PREPARATION OF STERILE DISCS.

(*i*) Press a sharp cork borer, 5 mm diameter bore, into a pile of Whatman No. 1 filter paper.

(*ii*) Autoclave the discs in a moisture-proof container at $103.4 \, kN/m^2$ for 15 minutes.

11.7.3 Isolation of *Streptococcus salivarius* from saliva

This bacterium is one of the commonest of the normal mouth flora. It is capable of hydrolysing sucrose into glucose and fructose and using the latter to form a slimy polysaccharide named levan. This is partly responsible for the formation of plaque on the teeth. Plaque is attacked by other bacteria to release acids which decay the tooth's structure.

1. Pour plates of sucrose agar.

2. Rub a sterile cotton wool swab (such as used for cleaning babies' ears and noses) against the inside of your cheek.

3. Swirl the end of the swab in a few cm^3 of sterile saline.

4. Inoculate the solidified sucrose agar plates with the saline using the streak method.

5. Incubate at 37°C for 3—4 days and look for colonies with a slimy, glassy appearance. Use these to prepare smears for Gram staining.

Questions

1. What is the appearance and Gram reaction of *S. salivarius*?

2. Devise an experiment to test the effect of toothpaste on this bacterium.

PREPARATION OF SUCROSE AGAR.

1. Dissolve 10 g of tryptone, 5 g yeast extract, 3 g of disodium hydrogen phosphate, 50 g sucrose and 15 g agar in 1 l distilled water.

2. Autoclave at $103.4 \, kN/m^2$ for 10 minutes and adjust pH to 7·4.

11.8 FOOD MICROBIOLOGY

11.8.1 The effect of pasteurization on milk

The method of counting bacteria in milk is very similar to that used for water (11.6.1).

1. Shake a 10 cm^3 sample of raw milk with 90 cm^3 of sterile saline to produce a dilution of 1 in 10 (10^{-1}). Use this to prepare dilutions of 10^{-2}, 10^{-3} and 10^{-4}.

2. Prepare pour plates of each dilution using milk agar (see 11.3.1 (C)) as the growth medium.

3. Invert the plates when the agar is solid and incubate at 30°C

for 3 days.

4. Pour a second 1 cm³ raw milk sample into a sterile test tube. Stopper it with a cotton wool bung.

5. Place the tube in a waterbath at $63.5 \pm 0.5°C$ for 35 minutes making sure that the milk is below waterlevel.

6. Remove the tube from the bath and use the milk to prepare dilutions of 10^{-1}, 10^{-2} and 10^{-3} with sterile saline solution.

7. Prepare pour plates for each dilution with milk agar and incubate at 30°C for 3 days.

8. Count the colonies on each plate and calculate the number of bacteria per cm³ of raw and pasteurized milk (see 11.6.1).

Questions

1. One method of pasteurizing milk is to hold it at 63°C for 30 minutes. Why was your sample placed in the water bath for 35 minutes?

2. Pasteurization should kill 99% of bacteria in raw milk. Do your results agree with this?

3. Why was milk agar used as the growth medium?

11.8.2 Microscopic examination of yoghurt

1. Place a drop of sterile water on a microscope slide. Inoculate with a loopful of natural yoghurt and make a thin heat fixed smear (see 10.5.1).

2. Stain the smear using Gram's method and examine the preparation under the microscope.

Questions

1. What are the shapes and Gram reactions of the most prevalent bacteria in the smear?

2. Which genera of bacteria are frequently used in the production of yoghurt?

3. What major chemical change do the yoghurt bacteria bring about in milk?

11.8.3 The effect of thawing frozen meat on microbial numbers

1. Thoroughly shake 10 g of just thawed minced beef in 90 cm³ of sterile saline. Use the liquid to prepare dilutions of 10^{-2}, 10^{-3} and 10^{-4} (see 11.6.1).

2. Prepare pour plates of each dilution using nutrient agar as the growth medium.

3. Invert the plates and incubate for 3 days at 30°C.

4. Allow a second 10 g of frozen minced beef to stand at room temperature for 24 hours, then use it to prepare dilutions and pour plates. Incubate the plates at 30°C for 3 days.

5. Without opening the plates count the colonies on each plate and calculate the number of bacteria per gram in each sample of minced beef (see 11.6.1).

Questions

1. From your observations what is the effect of allowing frozen food to stand for some time at room temperature?

2. Assuming that the bacteria have grown exponentially from the time the food thawed, calculate the generation time (see Chapter 7).

11.9 SOIL MICROBIOLOGY

11.9.1 Ammonification by soil microbes

1. Inoculate a tube containing 10–15 cm³ of 4% aqueous peptone solution with just enough soil to cover the tip of a sterile scalpel blade. Keep a second uninoculated tube as a control.

2. Incubate at 30°C for 3–4 days.

3. Place drops of Nessler's reagent on to a white tile and using a sterile pipette add a drop of the culture medium from each tube to the reagent. Mix with a glass rod. The presence of ammonia is indicated by the development of a yellow colour, the intensity of the colour being a measure of the activity of the ammonifiers. A deep brown colour denotes large amounts of ammonia.

4. Incubate the tubes further and repeat the test at 7–8 days.

Questions

1. Where does the ammonia produced in the cultures come from?
2. Which groups of soil microbes are responsible for the release of ammonia?
3. Where does ammonia come from in the soil environment?

11.9.2 Nitrification by soil microbes

A. CONVERSION OF AMMONIA TO NITRITE

1. Inoculate a flask containing $100 \, cm^3$ of ammonium sulphate medium with 1 g of soil. Set up a control with sterile autoclaved soil.
2. Incubate at 30°C for 7 days and aseptically remove $4 \, cm^3$ from both flasks. Test each sample separately as follows:
(i) Add an equal volume of 10% potassium iodide solution followed by a similar quantity of dilute sulphuric acid. Any nitrite present will oxidise the iodide to free iodine.
(ii) Add a few drops of starch solution to form the dark blue starch-iodine complex.

Questions

1. Name one genus of bacterium capable of converting ammonia to nitrite.
2. What benefit does the bacterium gain from the process?

PREPARATION OF AMMONIUM SULPHATE MEDIUM

1. Dissolve 2 g ammonium sulphate, 0.75 g dipotassium hydrogen phosphate, 0.25 g potassium dihydrogen phosphate, 0.01 g ferrous sulphate, 0.01 g manganese sulphate and 0.002 g of calcium chloride in 1 l distilled water.
2. Autoclave at $103.4 \, kN/m^2$ for 15 minutes.

B. CONVERSION OF NITRITE TO NITRATE

1. Inoculate a flask containing $100 \, cm^3$ of nitrite medium (recipe as for ammonia medium but $NaNO_2$ 2 g instead of $(NH_4)_2SO_4$) with 1 g of soil. Set up a control with sterile soil.
2. Incubate at 30°C for a week. Aseptically remove $3 \, cm^3$ of medium from both flasks.

Decompose any residual nitrite in each sample by adding 1 g of solid hydrazinium sulphate. Allow to stand for 5–10 minutes until effervescence ceases. Now test for nitrate by the brown-ring method:
(i) Add $3 \, cm^3$ of a freshly prepared saturated solution of ferrous sulphate.
(ii) Pour $3–5 \, cm^3$ of concentrated sulphuric acid (TAKE CARE!) slowly down the inside of the tube so that the acid forms a layer beneath the mixture. The formation of a brown ring within a few minutes where the two layers meet indicates the presence of nitrate.

Questions

1. Name one genus of bacterium capable of converting nitrite to nitrate.
2. What advantage does it obtain from this conversion?

149

11.9.3 Nitrogen fixation in soil

There is no elementary technique whereby it is possible to demonstrate the fixation of nitrogen gas but the organisms responsible can be isolated and examined.

A. SYMBIOTIC NITROGEN FIXERS

1. Remove one or two roots from a clover, lupin or field pea plant and gently wash off the soil with tap water.
2. Place small lengths of the root with obvious nodules into a small bottle of 0·1% mercuric chloride solution for 3 minutes.
3. Remove the root portions with sterile forceps and transfer them to a small bottle of 70% ethanol for 3 minutes.
4. Transfer the root material to a sterile bottle and wash thoroughly in several changes of sterile saline.
5. Using a sterile scalpel, crush one or two nodules in a few cm^3 of the saline.
6. Prepare smears and Gram stain samples of the crushed nodules.

Rhizobium is a Gram-negative, pleomorphic bacterium which appears as X, Y, T or even L-shaped forms.

B. NON-SYMBIOTIC NITROGEN FIXERS

Free living nitrogen fixing organisms, such as *Azotobacter*, can use mannitol as a carbon source but few other soil microbes can do so. As *Azotobacter* is aerobic it will grow on the surface of a solid medium.

1. Pour plates of Ashbey's nitrogen-free medium.
2. Sprinkle a few crumbs of soil on to the surface of the solidified medium.
3. Incubate at 25°C for a week by when creamy, glistening colonies should have developed.
4. Prepare smears of the growth obtained, Gram stain and examine microscopically.

Azotobacter spp. are Gram negative large rods.

PREPARATION OF ASHBEY'S NITROGEN-FREE MEDIUM

1. Dissolve 15 g mannitol, 0·2 g magnesium sulphate, 0·2 g dipotassium hydrogen phosphate, 0·2 g calcium chloride and 15 g of agar in 1 l distilled water. Add one drop of 10% solutions of ferric chloride and molybdenum trioxide.
2. Adjust the pH to 7·4 and autoclave at 103·4 kN/m^2 for 15 minutes.

C. ISOLATION OF *Clostridium pasteurianum*

This is the best known anaerobic nitrogen fixer. It is quite prevalent in many soils and may be more important than the aerobic species.

1. Inoculate 100 cm^3 of sterile nitrogen-free Winogradsky's medium. The medium is best placed in a 125 or 150 cm^3 flask in order to attain anaerobic conditions.
2. Incubate at 30°C for one week.
3. Examine the flask. Gas bubbles and a rancid smell are evidence of anaerobism.
4. Make Gram stained smears from samples taken from near the bottom of the medium.

C. pasteurianum is a Gram positive, sporing bacterium.

PREPARATION OF WINOGRADSKY'S MEDIUM

1. Dissolve 1 g dipotassium hydrogen phosphate, 0·2 g magnesium sulphate, 20 g glucose and 30 g of calcium

carbonate in 1 l of distilled water. Add one drop of 10% solutions of sodium chloride, ferrous sulphate and manganese sulphate.

2. Autoclave at 103·4 kN/m² for 10 minutes.

Questions

1. By what process do these bacteria fix atmospheric nitrogen?
2. What products are formed and what are they used for?
3. In what ways does *Rhizobium* depend on its leguminous host?
4. How do the free-living nitrogen-fixers obtain energy for synthetic processes?

11.10 MICROBIAL ECOLOGY

11.10.1 Fungal succession on herbivore dung

1. Incubate rabbit pellets in moist chambers at room temperature. Leave some of the chambers uncovered but cover others with lightproof paper leaving a small window in the middle of the lid.

2. Scan the dung at regular intervals over 2–3 weeks for the appearance of fungi, noting the time at which each species appears and for how long it remains active. Use flamed mounted needles to pick off specimens required for closer examination. Mount in a drop of lactophenol on a clean slide and squash under a cover slip.

3. Examine the window of the covered chambers for the presence of fungal spores.

Questions

1. How do you account for the presence of fungi in dung?
2. Which group of fungi sporulate first? Can you suggest a reason for this?
3. Why is sporulation of the other fungal classes somewhat later?
4. How do you account for the distribution of spores on the lids of boxes with windows?

MOIST CHAMBERS. These are clean transparent containers lined with filter or blotting paper moistened with 2% aqueous glycerol. Clear-plastic picnic boxes are ideal. The lids must be kept slightly open to allow adequate aeration. Kill any insects that appear with an aerosol insecticide. Sheep, cow or horse dung can be used but rabbit pellets are less offensive. For identification of fungi see Richardson and Watling (Appendix 3).

11.10.2 The colonization of buried Cellophane (after Tribe, 1967)

1. Boil pieces of plain Cellophane (grade P.T. 300) in distilled water for 15–20 minutes to remove plasticisers.

2. Cut into 1 cm × 0·5 cm rectangles and press one piece on to the centre of a clean microscope slide while the Cellophane is moist. Prepare six slides in this way.

3. Mark the back of the slide not used with a wax crayon.

4. Bury the slides vertically in 2 mm-sieved soil contained in a large beaker. Keep the soil moist throughout the investigation but do not flood it with water.

5. At weekly intervals remove one slide as shown in Fig. 11.1.

6. Wipe the back of the slide clean and carefully brush away any large particles of soil sticking to the Cellophane.

7. Add one drop of cotton blue in lactophenol followed by several drops of lactophenol.

8. Lower a large cover slip onto the Cellophane and examine the preparation under the microscope.

Fig. 11.1   Removing a Cellophane slide from soil.

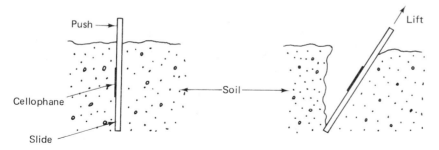

Questions

1. Which group of organisms first grows on to the Cellophane?
2. Suggest reasons for the success of this group as pioneer colonizers.
3. What is Cellophane composed of and what enzyme is required to hydrolyse it?
4. Do the initial colonizers change the microhabitat in any way and if so could the changes be advantageous to other groups of organisms?
5. Is your answer to 4 substantiated by later findings?
6. Are there any signs of animal activity during the course of the investigation? If so observe the feeding behaviour of these creatures and make annotated drawings to illustrate your findings.
7. Construct a simple food web from your observations over the period of the investigation.
EXTENSION WORK: The experiment can be extended by comparing the pattern of colonization in soils from different habitats such as garden, woodland or field soil, or the same soil may be kept under different regimes of pH, moisture content, fertilizer treatment, pesticide treatment and so on.

11.11 PATHOLOGY

11.11.1 Demonstration of Koch's postulates on soft fruit

In the late nineteenth century Robert Koch established a means of confirming that a particular organism was the causative agent of a specific disease. His procedure can be summarized in what are known as Koch's postulates:

(*i*) In all cases of the disease find the organism which is not present in healthy individuals.
(*ii*) Isolate the organism from the diseased hosts and grow it in pure culture.
(*iii*) Inoculate a healthy individual with the pure culture and observe whether or not the disease appears.
(*iv*) Re-isolate the organism from the host in which the disease has been induced and confirm that it is the same organism used in the inoculum.

The steps can be retraced as shown in Fig. 11.2. If there are difficulties in obtaining diseased fruit use pure cultures of known pathogens such as *Monilia fructigena*, *Penicillium expansum* or *Botrytis cinerea*.

Fig. 11.2 Demonstration of Koch's postulates.

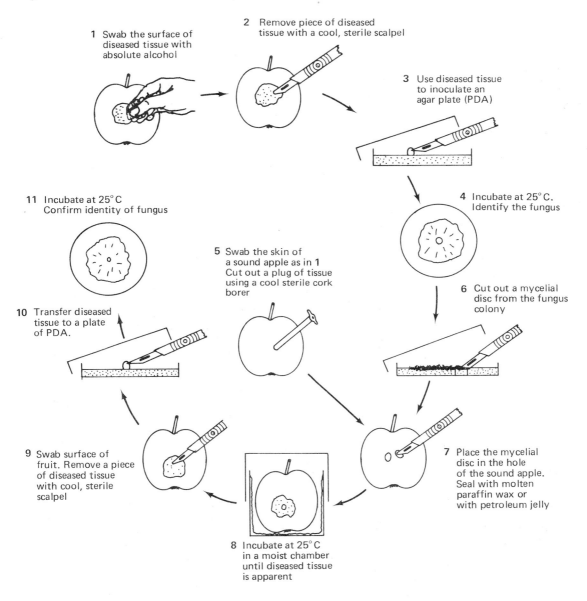

1 Swab the surface of diseased tissue with absolute alcohol

2 Remove piece of diseased tissue with a cool, sterile scalpel

3 Use diseased tissue to inoculate an agar plate (PDA)

4 Incubate at 25°C. Identify the fungus

11 Incubate at 25°C Confirm identity of fungus

5 Swab the skin of a sound apple as in 1 Cut out a plug of tissue using a cool sterile cork borer

6 Cut out a mycelial disc from the fungus colony

10 Transfer diseased tissue to a plate of PDA.

9 Swab surface of fruit. Remove a piece of diseased tissue with cool, sterile scalpel

7 Place the mycelial disc in the hole of the sound apple. Seal with molten paraffin wax or with petroleum jelly

8 Incubate at 25°C in a moist chamber until diseased tissue is apparent

Questions

1. Why is it necessary to swab the surface of the diseased fruit before removing some of the rotten tissue?
2. Why must the skin of the sound fruit be broken before it is inoculated with the fungus?
3. How would the fruit skin be damaged under natural conditions?
4. What enzymes do brown-rot fungi produce and what part do they play in the process of host colonization?

153

EXTENSION WORK: The investigation can be made more interesting by comparing the rate at which one species of fungus colonizes different varieties of apple or by comparing the growth rates of different species of fungi in the same variety of apple. Use the diameter of rotting tissue as indication of growth rate. Other species of fruit or vegetable can be used to demonstrate host specificity.

11.12 VIROLOGY

11.12.1 Demonstration of the presence of bacteriophages in soil

1. Inoculate 100 cm^3 of sterile nutrient broth in a 500 cm^3 conical flask with 5–10 g of soil.
2. Add 1 cm^3 of a 24-hour broth culture of *Escherichia coli*.
3. Incubate at 37°C for 24 hours then filter the soil culture through a bacteriological filter to obtain a 'phage-rich filtrate.
4. Prepare lawn plates of *E. coli* (see 10.3.1), allow to dry, then place spots of the 'phage-rich filtrate onto the surface.
5. Incubate the plates for 24–48 hours and examine for the presence of plaques. Do not open the plates when making these observations.

Questions

1. What is a plaque and what is the reason for its formation?
2. How would you modify the experiment to demonstrate the growth cycle of a 'phage?

11.13 PROTOZOA

11.13.1 Ingestion of food particles by *Paramecium*

1. Place a drop of a culture of *Paramecium* on to a cavity slide.
2. Add a drop of 10% methyl cellulose or 0·1% sodium alginate to slow down the movement of *Paramecium*.
3. Now add a drop of a suspension of yeast stained with Congo red.
4. Lower a cover slip into position and examine the preparation under the high power objective. Look for evidence of ingestion of the yeast and change in colour of the contents of food vacuoles.

Questions

1. How were yeast cells ingested by *Paramecium*?
2. Congo red is an indicator for the pH range of 3–5. At pH 3 it is blue-violet and at pH 5 it is red–orange. Using this information explain the changes occurring in the food vacuoles during the digestion of the yeast cells.
PREPARATION OF STAINED YEAST. Boil 3 g fresh yeast and 30 mg congo red in 10 cm^3 distilled water for 5–10 minutes. Allow to cool before use.

11.13.2 The effect of light on *Euglena*

1. Prepare a light-proof cover to fit around a tube culture of *Euglena* as shown in Fig. 11.3.
2. Leave the first hole uncovered, stick one piece of tracing paper over the second hole, two pieces over the third hole and three pieces over the fourth hole.
3. Stand the culture on a windowsill for a day with the holes in the light-proof cover facing towards the light. Notice the distribution of *Euglena* in the culture tube. Remove the cover and place the culture in a dark cupboard for 24 hours.
4. Replace the tracing paper with coloured cellophane; red over the first hole, yellow over the second, green over the third

and blue over the fourth.
5. Again stand the culture on a windowsill as in 3.
6. The following day again notice the distribution of *Euglena* in the culture tube.

Fig. 11.3 Cover for *Euglena* culture

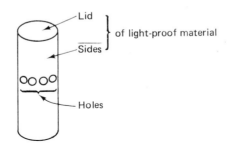

Lid
of light-proof material
Sides
Holes

Questions

1. How did *Euglena* react to
   (*a*) different intensities of light and
   (*b*) different wavelengths of light?
2. How does *Euglena* detect changes in light intensity?

11.14 FUNGAL GENETICS

11.14.1 Inheritance of spore colour in *Sordaria fimicola*

Although *Sordaria* is homothallic it will hybridize. The wild type strain has black ascospores but mutants with colourless spores are known. If these are grown together asci containing black and white spores are produced. An analysis of the number and distribution of the black and white spores will indicate the pattern of inheritance of the genes for spore colour.
1. Inoculate one end of a plate of corn meal agar (Oxoid Ltd.) with a mycelial disc of black-spored *Sordaria* and the opposite end with white-spored *Sordaria*.
2. Incubate for 7–10 days at 25°C.
3. Remove a number of the larger ascocarps from the centre of the dish where the growth of the two strains has overlapped.
4. Place the ascocarps on a microscope slide, add a drop of lactophenol and squash the fruiting bodies under a cover slip.
5. Examine your preparation under a microscope and draw as many different arrangements of spores as you can see in the hybrid asci.

Questions

1. When does nuclear fusion and meiosis occur in the life cycle of *Sordaria?* (see Chapter 6).
2. What is the proportion of black to white spores in the majority of the asci?
3. Mendel's first law of inheritance states:
   'Of a pair of contrasting characters only one is represented in each meiotic product.'
Does your answer to question 2 support this law?
4. In those asci with four white or four black spores at the tip, the genes for spore colour segregated at the first meiotic division. There was no crossing over between the locus for

155

spore colour and the centromere (Fig. 11.4). What would be the result if a chiasma formed between the locus and the centromere? (Fig. 11.5).

5. What would be the effect on the proportion of black and white spores if one of the genes for white spore colour mutated to the wild type gene during meiotic division? (Fig. 11.6). Is there any evidence for such a mutation in your observations? Conversely is there any evidence for the mutation of the wild type gene to a gene for white spore colour?

Figs. 11.4, 11.5, and 11.6   Inheritance of spore colour in *Sordaria*

11.4

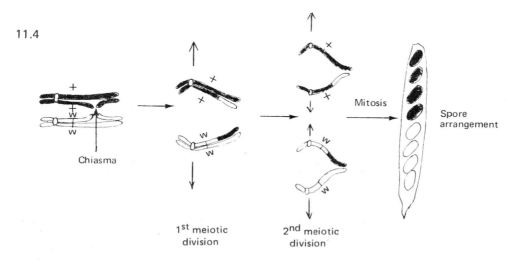

+ = gene for black spores, w = gene for white spores

What spore arrangement would occur if the two chromosomes were inverted during the first meiotic division?

11.5

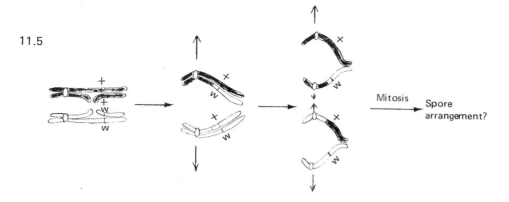

How would the spores be arranged if one of the homologous chromosomes was inverted during the second meiotic division?

11.6

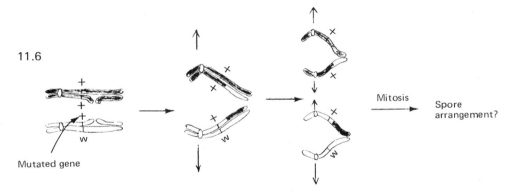

Mitosis → Spore arrangement?

Mutated gene

11.15 BLUE–GREEN ALGAE

11.15.1 Isolation of blue–green algae

Blooms of blue–green algae are sometimes seen as slimy masses in ponds, in slow-flowing streams, on the surface of water-logged soil, on damp brickwork and on pieces of unglazed clay plant pots left on the surface of damp soil. It is relatively easy to obtain cultures of blue–green algae from these habitats but the cultures are often contaminated with bacteria.
1. Place 125 cm³ of blue–green enrichment medium in sterile 250 cm³ conical flasks.
2. Add a small inoculum of soil, pond or stream water or of any habitat where blooms of blue-green algae are apparent.
3. Stopper the flasks with cotton wool bungs and incubate at 20–30°C in continuous light from a white fluorescent strip lamp or from a tungsten lamp. Place replicate flasks at different distances from the light source in order to establish the optimum light intensity for growth of the blue-green algae.
4. When the medium shows signs of turbidity remove small samples with a sterile pasteur pipette. Place on a clean slide, lower a cover slip in position and examine under a microscope.

Questions

1. Why are blue–green algae usually slimy?
2. Why is it necessary to provide light when incubating the cultures?
3. Suggest ways of preparing pure cultures of blue–green algae from your mixed cultures.

PREPARATION OF BLUE–GREEN ENRICHMENT MEDIUM:
1. Dissolve 0·2 g dipotassium hydrogen phosphate, 1 g ammonium nitrate, 0·1 g hydrated magnesium sulphate and 0·001 g ferrous chloride in 1 l of tap water.
2. Autoclave at 103·4 kN/m² for 15 minutes.

11.16 TEACHER'S NOTES FOR PRACTICAL EXERCISES

11.2 Question 2. The use of a number of different media and a range of incubation temperatures would enable a greater variety of contaminants to be isolated.
11.3.1 (D) Question 1. Tributyrin is an ester of glycerol and butyric acid.
11.3.2. *Aspergillus rugulosus* produces its own vitamins and secretes biotin into the medium. *Sordaria fimicola* grown in a biotin-deficient medium fails to produce asci and ascospores

in its perithecia. However if it is grown in dual culture with *A. rugulosus* it produces fertile perithecia even though the medium lacks biotin.

11.4.1 (B). Question 2. As oxidases catalyse the reduction of oxygen to water they are only found in aerobic bacteria.

11.4.1 (C). Question 2. Catalase activity results in the release of oxygen from hydrogen peroxide so that catalase positive bacteria must be aerobic.

11.4.3 (A). Question 3. Fermentation of one mole (180 g) of glucose yields two moles (2×22.4 l) of carbon dioxide:

$$C_6H_{12}O_6 \rightarrow 2CO_2 + 2C_2H_5OH$$

$$\therefore \frac{5}{180} \text{ moles yields } \frac{2 \times 22.4 \times 5}{180} \text{ l of } CO_2$$

$$= 1.25 \text{ l at STP}$$

The maximum increase in volume observed will be about 500 cm^3, thus the efficiency of the process is $\frac{500}{1250} \times 100 = 40\%$. The accumulation of alcohol is partly responsible for incomplete fermentation of the glucose.

11.5.1. It will not be possible to complete this exercise in a normal laboratory period. If readings cannot be taken at the suggested time intervals the cultures can be incubated in advance and stored in a refrigerator until required. Students can then count the cell numbers and calculate the cell densities during the time available.

11.6.1. Some bacteria will not grow on the medium used or at the temperature at which the plates are incubated. Thus a range of media and incubation temperatures would enable one to isolate a larger number of species.

11.7.2. A standard growth curve is constructed by plotting the diameter of the inhibition zone against the concentration of disinfectant.

Question 2. The rate of diffusion of the disinfectant through the agar is the most important factor and this will depend upon its solubility in the aqueous phase of the medium and on temperature.

11.7.3. Question 1. Gram positive cocci in long chains.

Question 2. A suitable method would be to prepare lawns or pour plates of *S. salivarius* using sucrose agar and then squeeze blobs of toothpaste on to the surface of the agar. Incubate and then examine for inhibition zones. The inhibitory effect of different brands of toothpaste can be tested in this way.

11.8.1. Question 1. The extra five minutes is needed to bring the tubes up to the pasteurization temperature.

11.10.1. Question 4. The spore-producing organs of some fungi are positively phototropic and release their spores towards a source of light.

11.11.1. Question 1. To remove superficial contaminants.

Question 4. Pectinases and cellulases help to separate the host cells thus facilitating growth of the fungal hyphae through the host's tissues.

APPENDIX 1

LIST OF SUPPLIERS

A.1.1 Equipment and reagents

1. Astell-Hearson Ltd., 172 Brownhill Road, Catford, London.
2. Baird & Tatlock Ltd., P.O. Box 1, Romford, Essex.
3. BDH Chemicals Ltd., Poole, Dorset.
4. Difco Laboratories, P.O. Box 14B, Central Avenue, West Molesey, Surrey.
5. Dyos Plastics, 242 Tolworth Rise South, Surbiton, Surrey.
6. A. Gallenkamp & Co. Ltd., P.O. Box 290, Technico House, Christopher Street, London.
7. Glaxo Laboratories Ltd., Greenford, Middlesex.
8. Grant Ltd., Barrington, Cambridge.
9. Gerrard & Haig Ltd., Gerrard House, Worthing Road, East Preston, Littlehampton, Sussex.
10. Griffin & George Ltd., 285 Ealing Road, Wembley, Middlesex.
11. Edward Gurr Ltd., Michrome Laboratories, 42 Upper Richmond Road West, London.
12. Gelman Hawksley Ltd., 12 Peter Road, Lancing, Sussex.
13. Hopkin & Williams Ltd., Freshwater Road, Chadwell Heath, Essex.
14. Koch-Light Laboratories, Colnbrook, Bucks.
15. Millipore Ltd., Millipore House, Abbey Road, London N.W. 10.
16. Oxoid Ltd., Wade Road, Basingstoke, Hants.
17. D. J. Parry Ltd., Avonmore Road, Hammersmith, London W.14.
18. Philip Harris Biological Ltd.
19. Sigma Chemical Co. Ltd., Norbiton Station Yard, Kingston-upon-Thames, Surrey.
20. Sterilin Ltd., 12 Hill Rise, Richmond, Surrey.

Equipment and reagents supplied:	Suppliers:
Autoclaveable bags	18, 20
Autoclaves	1, 2, 6, 9, 10, 18.
Antibiotics	7, 18.
Cellophane	17, 18.
Cellulose powder	4, 18, 19.
Chemicals and stains	3, 11, 13, 18, 19.
Filters (bacteriological)	15, 18.
General equipment (inc. glassware)	2, 6, 9, 10, 18.
Haemacytometers	6, 10, 12, 18.
Incubators	1, 2, 6, 10, 18.
Inoculating cabinets	18.
Media	4, 16, 18.
Micrometers	6, 10, 18.
Multodisks	16, 18.

Petri dishes (plastic)	5, 18, 20.
Water baths	6, 8, 10, 18.
Wires, loops and spreaders	1, 10, 18.

A.1.2 Microbial strains

(*i*) *Commercial suppliers:*
Gerrard & Haig Ltd.
Philip Harris Biological Ltd.
Oxoid Ltd.
(*ii*) *Other sources:*
ALGAE AND PROTOZOA:
Culture Collection of Algae and Protozoa, Botany School, University of Cambridge, Cambridge.
BACTERIA:
National Collection of Industrial Bacteria, Torry Research Station, P.O. Box 31, 135 Abbey Road, Aberdeen.
National Collection of Dairy Organisms, National Institute for Research in Dairying, Shinfield, Reading.
FUNGI:
Commonwealth Mycological Institute, Ferry Lane, Kew, Surrey.
National Collection of Yeast Cultures, Brewing Industry Research Foundation, Lyttel Hall, Nutfield, Nr. Redhill.

Most of the apparatus, materials and organisms are available from Biological Supply Houses. Philip Harris Limited and Philip Harris Biological Limited have undertaken to cover all requirements and to issue appropriate lists.

APPENDIX 2

LIST OF CULTURES USED IN THE PRACTICAL EXERCISES

BACTERIA:
Aerobacter aerogenes
Azotobacter agilis
Bacillus cereus
B. subtilis
B. stearothermophilus
Escherichia coli
Lactobacillus acidophilus
L. lactis
Micrococcus agilis
M. luteus
Mycobacterium phlei
Proteus vulgaris
Pseudomonas fluorescens
Spirillum serpens
Staphylococcus albus
Streptococcus faecalis

FUNGI:
Aspergillus repens
A. rugulosus
Aspergillus sp.
Botrytis cinerea
Chaetomium sp.
Fusarium sp.
Monilia fructigena
Mucor sp.
Myrothecium verrucaria
Penicillium expansum
P. wortmanni
Penicillium sp.
Saccharomyces cerevisiae
S. cerevisiae var. ellipsoideus
Sordaria fimicola
Trichoderma sp.

161

APPENDIX 3

REFERENCES

Bottle, R. T. (1967). 'Zero order yeast fermentation.' *Education in Chemistry, 4,* 197–9.

Dade, H. A. and Gunnell, J. (1969). *Classwork with Fungi.* 2nd edition. Commonwealth Mycological Institute.

Dixon, A. F. G. (1973). 'Biology of Aphids.' (*Institute of Biology Studies in Biology No. 44*). Arnold.

Gilbert, O. L. (1969). 'The effect of sulphur dioxide on lichens and bryophytes.' *European Symposium on the influences of Air Pollution on Plants and Animals.* Wageningen, Netherlands.

Gilbert, O. L. (1970). 'Further studies on the effects of sulphur dioxide on lichens and bryophytes.' *New Phytologist, 69,* 605–28.

Harley, J. L. (1971). 'Mycorrhiza.' *Oxford Biology Readers No. 12.* Oxford University Press.

Hatch, A. B. (1937). 'The physical basis of mycotrophy in the genus *Pinus.*' *Black Rock Forest Bulletin, 6, 168 pp.*

Richardson, M. and Watling, R. Keys to fungi on dung. *Bulletin of the British Mycological Society, 2* (1), 18–43 (1968) and **3** (2), 86–124 (1969).

Stanier, R. Y. and van Niel, C. B. (1962). 'The Concept of a Bacterium.' *Arch. Mikrobiol., 42,* 17–35.

Tribe, H. T. (1967). 'Practical studies on the biological decomposition of cellulose.' *School Science Review, 49,* 95–112.

Trinci, A. P. J. (1972). 'Culture turbidity as a measure of mould growth.' *Transactions of the British Mycological Society, 58,* 467–73.

Watson, M. A. (1967). 'Epidemiology of aphid-transmitted plant diseases.' *Outlook on Agriculture, 5,* 155–66.

APPENDIX 4

SUGGESTIONS FOR FURTHER READING

A.4.1 Theoretical texts

Alexopoulos, C. J. (1962). *Introductory Mycology*. J. Wiley and Sons.

Brian, P. W. (1972). 'The economic value of fungi.' *Transactions of the British Mycological Society*, **58,** 359–75.

Brock, T. D. (1966). *Principles of Microbial Ecology*. Prentice-Hall.

Casida, L. E. Jr. (1964). *Industrial Microbiology*. J. Wiley and Sons.

Deverall, B. (1971). *Fungal Parasitism*. (Institute of Biology Studies in Biology No. 17). Arnold.

Frazier, W. C. (1967). *Food Microbiology*. 2nd edition. McGraw-Hill.

Gray, T. R. G. and Williams, S. T. (1971). *Soil Micro-organisms*. Oliver and Boyd.

Gillies, R. R. and Dodds, T. C. (1965). *Bacteriology Illustrated*. E. and S. Livingstone.

Hawker, L. E. and Linton, H. H. (1971). *Micro-organisms. Function, Form and Environment*. Arnold.

Hudson, H. J. (1972). *Fungal Saprophytism* (Institute of Biology Studies in Biology No. 32). Arnold.

Ingold, C. T. (1973). *The Biology of Fungi*. 2nd edition. Hutchinson.

Pelczar, M. J. and Reid, R. D. (1972). *Microbiology*. McGraw-Hill.

Scott, G. D. (1969) *Plant Symbiosis* (Institute of Biology Studies in Biology No. 16). Arnold.

Stanier, R. Y., Doudoroff, H. and Adelberg, E. A. (1971). *General Microbiology*. 3rd edition. Macmillan.

Smith, D. C. (1973). *The Lichen Symbiosis*. Oxford Biology Readers No. 42. Oxford University Press.

Smith, K. M. (1962). *Viruses*. Cambridge University Press.

Taber, W. A. and Taber, R. A. (1969). *The Impact of Fungi on Man*. J. Murray.

Vickerman, K. and Cox, F. E. G. (1967). *The Protozoa*. J. Murray.

Webster, J. (1970). *Introduction to Fungi*. Cambridge University Press.

A.4.2 Practical manuals

Alexopoulos, C. J. and Beneke, E. S. (1962). *Laboratory Manual for Introductory Mycology*. Burgess Publishing Co.

Dade, H. A. and Gunnell, J. (1969). *Classwork with Fungi*. Commonwealth Mycological Institute.

Harrigan, W. F. and McCance, M. E. (1966). *Laboratory Methods in Microbiology*. Academic Press.

Millipore. (1969). *Experiments in Microbiology.* Millipore Corporation.

Oxoid Manual (1969). 3rd edition. Oxoid Ltd.

Pawsey, R. K. (1974). *Techniques with Bacteria.* Hutchinson.

Pelczar, M. J. and Chan, E. C. S. (1965). *Laboratory Exercises in Microbiology.* McGraw-Hill.

Pramer, D. and Schmidt, E. L. (1964). *Experimental Soil Microbiology.* Burgess.

Seeley, H. W. and Vandermark, P. J. (1962). *Microbes in Action.* Freeman.

Smith, G. (1960). *An introduction to Industrial Mycology.* 6th edition. Arnold.

Weiser, H. H. (1962). *Practical Food Microbiology and Technology.* Avi Publishing Co. Inc.

APPENDIX 5

1. Write a general account of the characteristics of bacteria, with particular reference to their importance as agents of disease and decay.

2. Make a labelled diagram to show the ultra-structure of a typical bacterial cell as revealed by the electron microscope. Explain the meanings of the terms autotrophic and chemo-synthetic by reference to named bacteria.

3. Devise an experiment to show the effect on the growth rate of a bacterium of one of the following: nutrients, temperature, pH, oxygen availability. What results would you expect and what controls would you use to establish that the results were not due to a change in the organism itself during the course of the experiment?

(JMB. Specimen question for micro-organisms option.)

4. Give an account of the structure and chemical composition of viruses. In what respects do viruses differ from bacteria?

5. Write a general account of viruses, with particular reference to them as agents of plant and animal diseases.

6. For each of the following problems, devise an experiment to test the hypothesis or assumption expressed in it, making use of the following divisions:
(a) A concise explanation of the principles on which the experiment is based.
(b) An account of the procedures to be used. You may use simple flow diagrams to illustrate the relationships between parts of the experiment.
(c) An explanation of how the results would be presented and analysed.
(i) A dense culture of bacteria, which was turbid, became clear when the bacterial cells died and disintegrated. Two hypotheses were put forward to explain the death of the bacteria:
(a) they had been attacked and killed by a virus;
(b) they had been killed by a poisonous material accidentally included in the culture medium.
Devise an experiment to decide which hypothesis was false, bearing in mind that viruses can reproduce themselves and that poisons do not.
(ii) Vitamin B12 is a growth factor required by the unicellular organism *Euglena gracilis*. In pernicious anaemia in man, the body is deficient in vitamin B12. Given separate samples of serum from two different patients suffering from pernicious anaemia how could you use *Euglena* to compare the extent to which each was deficient in vitamin B12?

(Nuffield 'A' level Biology 1.2 (1970) and 1.2 (1971)).

7. Three bacterial mutants are unable to grow on a medium lacking substance Z. The normal wild type of bacterium synthesizes Z from inorganic materials in three stages, each of which requires a specific enzyme. The normal biochemical sequence leading to the production of Z is shown below.

$$\text{Inorganic materials and energy source} \xrightarrow[\text{enzyme A}]{} \text{substance X} \xrightarrow[\text{enzyme B}]{} \text{substance Y} \xrightarrow[\text{enzyme C}]{} \text{substance Z}$$

Pure strains of the mutants were streaked on to an agar plate containing inorganic materials and a substance which provided an energy source. After incubation the plate had the appearance shown in Fig. A.5.1. Breeding tests showed that each mutant strain lacked the ability to produce Z because it lacked one of the three necessary enzymes in the biochemical sequence, a different one in each case. Any substance which could be made, but not used, accumulated and diffused into the agar around the bacterial colonies. Mutant 2 was unable to produce enzyme A. Other tests showed that none of these strains produced an inhibiting substance which could stop the growth of the others.

Fig. A.5.1

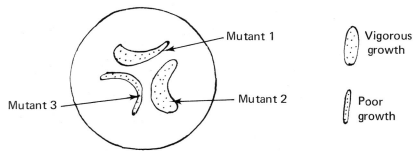

(a) Mutant 2 lacked enzyme A and could not make substance X from raw materials:
(i) From which two substances could mutant 2 make Z?
(ii) What could be the origin of these two substances?
(b) Explain the following:
(i) Mutant 1 had little growth next to mutant 2.
(ii) Mutant 1 had vigorous growth next to mutant 3.
(c) Which enzyme was lacking in mutant 3?
(d) How would you test your answer to (c)?
(Nuffield 'A' level Biology 2.1 (1969).)

8. Give a reasoned classification of the protozoans you have studied. What common characters have they to justify grouping them together as Protozoa? Justify carefully and fully each subdivision you describe. What useful purposes are served by such a classification?

(JMB. 'A' level Zoology 2 (1967).)

9. Compare the structural and functional organization of *Amoeba* with that of *Paramecium*. To what extent can these animals be regarded as equivalent to a single cell of a metazoan animal?

(JMB, 'A' level Zoology 2 (1968).)

10. Describe the locomotion of *Amoeba, Paramecium* and *Euglena*. Discuss how far locomotion and feeding are inter-dependent in these three protozoans.

11. By reference to the malarial parasite, discuss those features shown by the life-cycle which may be associated with its parasitic way of life.

12. By means of a clearly labelled diagram, outline the life-cycle of a named sporozoan parasite. Discuss the methods of control available against the parasite, relating them to the life-cycle where possible.

13. A protozoan species was cultured alone in (*i*) a sterile medium and (*ii*) a sterile medium to which was added a sterile extract from the medium in which another protozoan species had been cultured. A colorimeter was used to measure the growth of the protozoan cultures in both the sterile medium and in the medium with added extract, optical density (OD) readings being taken every two days. The results are shown on the graph in Fig. A.5.2—optical density curves of protozoan cultures grown in a sterile medium (S) and in a medium with added extract (E).
(*a*) Through which culture was more light transmitted?
(*b*) Give two precautions which would be necessary when taking the OD measurements.
(*c*) What is the effect of the extract on the growth of the protozoan species under test?
(*d*) Suggest one hypothesis to explain the results shown on the graph.

(Nuffield 'A' level Biology 2.1 (1970).)

Fig. A.5.2

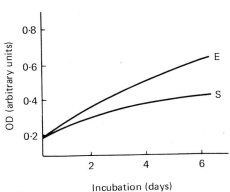

14. What are the major similarities and differences between the cellular anatomy of either bacteria or fungi and that of the cells of higher organisms?
(JMB. Specimen questions for micro-organisms option.)

15. For those fungi you have studied compare the production and liberation of spores.

16. When '+' and '−' strains of *Mucor* were grown together zygospores were formed. The zygospores were germinated and gave rise to germ sporangia. When the spores from these sporangia were tested it was found that all sporangia formed spores which were either all '+' or all '−'. Give an illustrated account of the morphological and cytological phenomena associated with zygospore formation and germination which explains the behaviour of the germ sporangia.

(JMB. 'A' level Botany 1 (1964).)

17. Write brief accounts of the following: alcoholic fermentation, mycorrhiza, fungal mycelia and heterothallism.

18. (*a*) What is understood by the term heterotrophic nutrition?
    (*b*) What structural and physiological features of a named fungus may be related to this form of nutrition?
    (*c*) By reference to examples give three different ways in which fungi are (*i*) beneficial to man, and (*ii*) harmful to man.

(J.M.B. 'A' level Biology 1 (1970).)

19. With reference to fungi, distinguish between parasites, saprophytes and symbionts. Give an account of the life-history of any named parasitic fungus. Why is this fungus considered to be a nuisance? State how its effects can be controlled.

(JMB. 'A' level Botany 2 (1968).)

20. Give an illustrated account of those features of the structure and life-history of a named parasitic fungus which are considered to be adaptations to a parasitic way of life. How does the relationship between a mycorrhizal fungus and the roots of a tree differ from that between a parasite and its host?

21. A dead plant is found to contain fungal mycelia. What procedure could be adopted to determine whether the fungus was a parasite which had brought about the death of the plant or a saprophyte which had invaded the tissues after death, or whether both a parasite and saprophyte were present?

(JMB. 'A' level Botany S (1957).)

22. Explain how you would obtain a pure culture of a fungus such as *Mucor* from some mouldy bread. What are the chief growth requirements of such a saprophytic fungus? How does the fungus utilize the substances it absorbs?

23. Write an account of symbiotic relationships between the roots of seed-bearing plants and other organisms.

24. *Pinus* seedlings were grown in sterilized soil, and one group of seedlings was inoculated with a fungus resulting in the formation of mycorrhiza, whilst the other group was not inoculated. When the two groups of seedlings were compared some months later the following results were obtained:

	Dry weight (mg)	% Dry weight		
		Nitrogen	Phosphate	Potassium
Mycorrhizal seedlings	404·6	1·24	0·20	0·74
Non-mycorrhizal seedlings	320·7	0·85	0·08	0·43

State what you conclude from these results and how these conclusions can be explained.

(JMB. 'A' level Botany 1 (1960).)

25. Yeasts and many other organisms respire aerobically and anaerobically. Explain the essential differences between these two processes. Briefly describe an experimental method by which you could demonstrate each of these processes in any one named organism.

(JMB. 'A' level Botany 2 (1968).)

26. Samples of sterile dung were inoculated with suspensions of fungal spores which were pretreated with alkaline pancreatin at 37°C. Data on the fungi are given in the table below.

Fungus	Latent period before spore germination starts (hr)	Growth rate of mycelium (mm hr^{-1})	Days after germination before fruiting bodies appear
Phycomycetes			
Species P1	5–8	9·1	2
Species P2	6–9	4·8	4
Ascomycetes			
Species A1	6–8	1·8	6
Species A2	4–6	19·0	9
Basidiomycetes			
Species B1	5–8	3·2	11
Species B2	10–12	3·7	37

(i) Suggest why the spores were pretreated with alkaline pancreatin at 37°C.

(ii) List the carbon sources of dung in the order in which you would expect them to be utilized by the fungi growing in it. Give your reasons.

(iii) Which of the fungi in the table would you expect to use most of the readily-utilizable carbon source? Give your reason.

(iv) Two explanations for the succession in the appearance of fruiting bodies have been suggested. The first is that each fungus has a characteristic time to fruit. The second is based on nutrition and states that the succession in the appearance of fruit bodies reflects the sequence of utilization of the carbon sources by the dung fungi. In general, the *Phycomycetes* have neither cellulase nor ligninase, the *Ascomycetes* have cellulase and the *Basidiomycetes* have both of these enzymes. Discuss which of the two explanations best fits the data given above.

(JMB. 'A' level Biology 2 (1974), (modified).)

27. A test tube of sterile nutrient medium was inoculated with the yeast *Schizosaccharomyces pombe* and kept at a constant optimum temperature for 24 hours. At intervals, the number of

cells present was estimated, using a counting chamber. The diagrams in Fig. A.5.3 show the appearance of a portion of the counting chamber with samples taken at different times after inoculation.

Fig. A.5.3

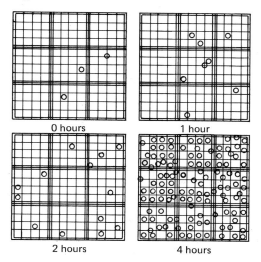

0 hours	1 hour
2 hours	4 hours

(*a*) (*i*) Use the information from the diagrams to complete the table below:

Hours after inoculation	0	1	2	4	5	8	15
Number of cells counted in the same area size as shown in the diagram					240	360	400

(*ii*) Construct a graph of these results.
(*b*) Briefly state what is happening at each of the distinct stages of population growth shown by the graph.
(*c*) What difference in the result would you expect
    (*i*) if the population had been grown at a temperature 10°C lower than that used in the experiment?
    (*ii*) if the concentration of the nutrient medium were doubled?

(Nuffield 'A' level Biology 2.1 (1973).)

28. (*a*) Discuss the part played by micro-organisms in the breakdown of plant material.
    (*b*) Cotton fibres consist almost entirely of cellulose.
When cotton fabrics are attacked by a cellulolytic fungus, the enzyme cellulase, secreted by the fungus, causes loss in mechanical strength of the fabric. Experiments were carried out on the cellulolytic fungus *Myrothecium verrucaria* as follows:
The first experiment was to determine the effect of sucrose concentration on the rate of growth and was done by estimating the dry weight of mycelium of the fungus at intervals after inoculating flasks containing mineral salts with either 1·0% or 10·0% of sucrose. The result of the experiment is shown in

Fig. A.5.4. The second experiment was to determine the effect of sucrose concentration on the cellulolytic activity of the fungus. This was done by inoculating strips of cotton fabric with fungus spores and suspending the inoculated fabric in the same mineral solution with 0·0%, 1·0% or 10·0% sucrose. The mechanical strength of the fabric was tested at intervals of time by applying increasing tension until the fabric broke. The results are expressed in Fig. A.5.5 as residual cloth strength as a percentage of the original value for uninoculated cloth. Examine the graphs and then answer the following questions:

Fig. A.5.4.

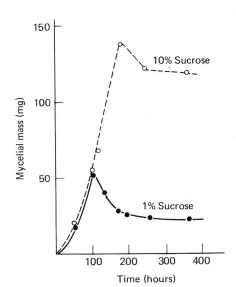

Fig. A.5.5.

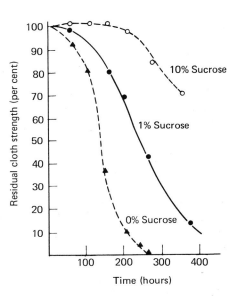

(*i*) Why do you think the fungus continues to increase in dry weight on 10·0% sucrose for a longer time than on 1·0% sucrose?

(*ii*) Why do you think the maximum dry weight reached on 10·0% sucrose is only about three times as much as on 1·0% sucrose?

(*iii*) What explanation can you suggest for the decrease in dry weight after the maximum value was reached?

(*iv*) Why do you think the slopes of the curves in Fig. A.5.5 are so different?

(*v*) In Fig. A.5.5 there is no appreciable decrease in strength of the cloth until about 160 hours in the absence of sucrose. Can you suggest an explanation?

(*vi*) If glucose had been substituted for sucrose in the second experiment, would you have expected similar results? Give reasons for your answer.

(*vii*) Do you think that the enzyme cellulase is likely to have any effect on sucrose?

(*viii*) Briefly comment on anything that strikes you as interesting or surprising about these results.

(JMB. Specimen question for micro-organisms option.)

29. Domestic sewage is rich in saprophytic bacteria and in organic material which is readily decomposed. When such sewage is discharged into a river it causes a number of chemical and biotic changes for some distance downstream from the 'out-fall' (i.e. the point at which it enters the river). Typical changes have been summarised by Hynes (1960) in the form of graphs in Fig. A.5.6.

Give reasoned explanations for the changes expressed by the curves, relating them to each other where you are able.

Fig. A.5.6.

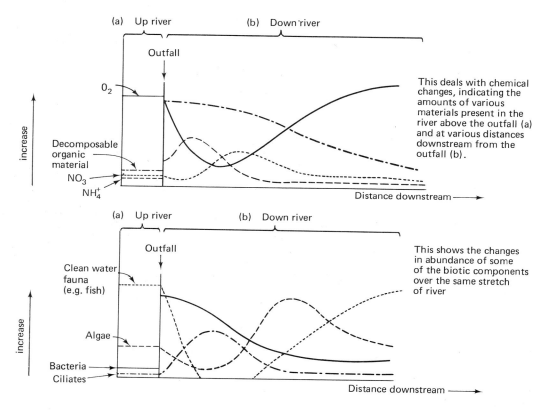

(JMB. 'A' level Biology 2 (1971).)

30. The fundamental requirements of sewage disposal are to remove objectionable solids and to lower the biochemical oxygen demand (BOD) before the sewage remains are discharged into water. Say, briefly, why these two requirements must be met.

Explain fully how micro-organisms in the sewage treatment plant accomplish the work of rendering sewage fit to be released into surface water.

(JMB. Specimen question on micro-organisms option.)

31. Beadle and Tatum used two different strains of the fungus *Neurospora* when investigating the relationship between genes and enzymes. A cross was made between a normal strain and a strain requiring vitamin B6. Each spore from the groups of eight formed in ascus sacs later in the life cycle of the hybrid was isolated and grown on a medium lacking vitamin B6. One set of results is shown below.

Spore number	Ascus number 1	2	3	4	5	6
1	—	—	—	—	—	N
2	s	—	s	—	—	N
3	s	N	—	N	—	N
4	s	N	—	—	—	N
5	N	—	—	—	N	s
6	N	—	—	—	N	s
7	N	s	—	—	s	s
8	—	s	N	s	s	s

N is normal growth on medium without vitamin B6.
s is slight growth on medium without vitamin B6.
— is spore failed to germinate.

The spores of *Neurospora* are produced as shown in Fig. A.5.7.
(a) If the wild-type gene is represented by + and the mutant (i.e. B6-requiring) gene by b, what would be the genotype of the diploid mother cell?
(b) What would be the genotypes of the four cells in C?
(c) Beadle and Tatum wrote about their results: 'It is clear from these rather limited data that this inability to synthesize vitamin B6 is transmitted as it should be if it were differentiated from normal by a single gene.' What information in the table shows why this conclusion is justified?

(d) Account for the difference in sequence between the spores in ascus number 5 and ascus number 6 shown in the table.

Fig. A. 5.7.

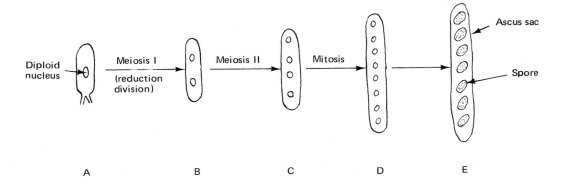

(Nuffield 'A' level Biology 2.1 (1971).)

INDEX

Bold figures refer to practical section.

Phytoalexins, 78
Phytomastigina, 37
Phytophthora infestans, 58, 64, 77
    control of, 111–12
    life-cycle of, 110–11
Plaques, 50, 53–4, **154**
Plasmodium, 43–4
    control of, 44, 46
    life-cycle of, 44–5
Pleomorphy, 15
Poliomyelitis, 109
    virus, 50
Pollution, effect on lichens, 82
Potato blight (see *Phytophthora infestans*)
Pour plates, **124**
PPLO (see *Mycoplasmatales*)
Preservation of food, 101–2
Presumptive coliform test, **145–6**
Prokaryota, 11
Promycelium, 67–8
Protista, 11
Protein levels, of microbes, 99
Protozoa, classification of, 33
    Amoeba, 34–7
    Euglena, 37–40
    Paramecium, 40–3
    Plasmodium, 43–6
Provirus, 54–5
Pseudomonadales, 13
Psychrophilic, 87
Puccinia graminis, 57, 65–6
Purification, of viruses, 50–1
    of water, 105–6

Replication of, viruses, 51–3
Reproduction, of bacteria, 22–3
    of blue–green algae, 32
    of fungi, 63–71
    of lichens, 81–2
    of *Protozoa*, 36–7, 40, 42–3, 44
Respiration, 25, 88
    respiratory enzymes, 88, 140–1
Rhizomorphs, 72–3
Rhizopoda (see *Sarcodina*)
Root nodules, 27

Saccharomyces, 58
    budding of, 64–5

structure of, 60
Saccharomyces cerevisiae, fermentation by, 92–3, **142**
    growth curve of, **143**
    life-cycle of, 68
    sporulation of, **122**
Saccharomyces cerevisiae var. ellipsoideus, 91, **143**
Saccharomyces carlsbergensis, 92
Safety procedures, 116–17
Safety references, 117
Salmonella, 101–3
Salmonella paratyphi, 107
Salmonella typhi, 107–8
Sap inoculation, 51
Saprolegnia, 56, 58, 63–4, 67–8
Saprophytes, 75–7
Sarcodina, 33–4
Sclerotia, 71–2
Schizogony, 44
Schizomycetes (see Bacteria)
Schizosaccharomyces, 56
Serology, 107–8
Sewage, treatment of, 103–5
Sordaria, 56, 58
Sordaria fimicola, genetics of, **155–6**
    life cycle of, 69
Soredia, 81
Spirochaetales, 14
Spores, of fungi,
    aplanospores, 64
    ascospores, 56, 68–9
    basidiospores, 57, 70–1
    chlamydospores, 72
    conidia, 56–7, 64–6
    oospores, 68, 111
    teleutospores, 57
    uredospores, 65–6
    zoospores, 63
    zygospores, 67–8
Sporogony, 44
Sporozoa, 33, 43
Staining, of bacteria, **128–31**
    of fungi, **131–2**
Stock cultures, maintenance of, **126–7**
Streak plates, **123–4**
Streptococcus salivarius, isolation of, **147**
Streptomycin, 95–7
Streptomyces griseus, 96

Streptomyces venezualae, 97
Sulphur cycle, 28–9
Symbiosis, 27, 78

Temperate 'phage, 54
Temperature, effect on fungal growth, **120, 143–4**
    effect on microbial growth, 87
Thermophilic, 87
Tissue culture, 50–1
Tobacco mosaic virus, 48–50
Toxins, bacterial, 103
    fungal (see aflatoxins)
Trace elements, 74
Transduction, 22, 55
Transformation, 22
Typhoid fever, control of, 108–9
    transmission of, 107–8

Vaccination, 109
Vaccines, 50, 109
Vinegar (see Acetic acid production)
Viruses,
    cultivation of, 50–1
    growth curve of, 53–4
    replication of, 51–2
    size range of, 50
    structure of, 47–9
Vitamin, assay by microbes, 89
    production by fungi, 75, **139**
    requirements of *Euglena*, 39
    requirements of fungi, 74–5

Water, number of bacteria in, 105, **145**
    purification of, 105–6
Wine, production of, 91–2

Yeast (see *Saccharomyces*)
Yoghurt, 99
    microscopic examination of, **148**

Zoomastigina, 37
Zygomycotina, 58

̇turn on or before the